D1121019

Advance Praise

THE HOLY UNIVERSE

"For those of us who have respect for the scientific method—but who also call ourselves 'spiritual but not religious,' who seek a spirituality unburdened by the chains of dogma, who seek a deep connection with Creation that is based in awe and reverence for nature and the cosmos—David Christopher has done a marvelous job in helping us reclaim the word 'Holy.'"

> —JOHN ROBBINS, *New York Times* bestselling author of
> *Diet for a New America, The Food Revolution,* and others

"Insightful and thought-provoking; connects the dots between challenges faced by ancient cultures and those confronting us today, in ways that inspire us all to take action."

> —JOHN PERKINS, *New York Times* bestselling author of *Hoodwinked,*
> *Confessions of an Economic Hit Man, Psychonavigation,* and others

"The story of our shared origins—not just as humanity or even life, but as the Universe itself—must be told again and again for us to come home once more in time and space, to be at ease and connected. David's telling as a conversation between Sage and Seeker brings the vastness of the story closer to our daily experience. His is a fresh and accessible voice."

> —VICKI ROBIN, *New York Times* bestselling coauthor of
> *Your Money or Your Life,* author of *Blessing the Hands That Feed Us*

"From past to present, from mythos to mapping a practical path, *The Holy Universe* offers powerful insights, inspiring us all to care—and then take conscious action from a deeply connected place."

> —JULIA BUTTERFLY HILL, activist, and author of
> *The Legacy of Luna, One Makes the Difference,* and
> *Becoming: Pictures, Poems, and Stories*

"David Christopher has done a brilliant job of explaining the new science of the evolution of life and the Universe. His story helps us view the crises of our time in a new and more positive light and offers a guidepost for the emerging culture of our time."

—PAUL H. RAY, PH.D., coauthor of
The Cultural Creatives: How 50 Million People Are Changing the World

"Truly remarkable—unites a clarity of understanding of what the human family is up against with a poetic story of where we came from and who we really are. David is an artist and an activist, a poet and a prophet. His wisdom is vibrant, moving, and powerful."

—LYNNE TWIST, cofounder, The Pachamama Alliance, founder,
The Soul of Money Institute, and author of *The Soul of Money*

"This book is a remarkable synthesis of modern and ancient wisdom, innovation, and science. David Christopher sheds light on the creative third options that emerge when polarities or opposites are brought together to create a greater whole. Timeless, yet most relevant for our current issues and challenges we face personally and collectively."

—ANGELES ARRIEN, PH.D., Cultural Anthropologist,
author of *The Four-Fold Way, Living in Gratitude,* and others

"*The Holy Universe* is a surprising and stunningly coherent weaving and celebration of both the Universe's magnificence and the healing possibilities embedded in our unfolding cosmic journey."

—BELVIE ROOKS, cofounder, Growing a Global Heart

"This touching story is instructive and hopeful, two very important ingredients for these times. It both reinforced my hope that the Sages of the world are there to help those of us who seek and that our lives are never that far off track—righted by a new perspective, a new possibility, just like those described in this heart-warming tale."

—CAROLYNE STAYTON, Executive Director, Transition US

"As I stand in the radiance of our holy Universe and peer into the gospel of the stars, it awakens the ancient sermon that lies deep within my soul. *The Holy Universe* is profoundly evocative of that sermon, and David Christopher is a modern-day sage and a master of the Coyote teaching tradition."

—DEDAN GILLS, poet

"A great fable! A story about story, and what games we humans play on ourselves . . . creating separation and fear out of our imaginations, wasting our creative skills while distancing ourselves from one another and Creation. *The Holy Universe* serves as a roadmap to return to the wisdom of the ancients, before we tricked ourselves into thinking things are different from how they really are."

—JOHN RENESCH, global futurist, and author of *Getting to the Better Future: A Matter of Conscious Choosing* and others

"This is a beautifully written work about the majestic Universe of which we are a part and to which we respond with awe and reverence."

—DR. LOWELL GUSTAFSON, PH.D., Secretary, International Big History Association; Professor of Political Science, Villanova University

"David Christopher creates a much-needed bridge between a sense of cosmic wonder and our everyday experience, showing how our understanding of the meaningful unfolding of the Universe can actually inform our day-to-day struggles, questions, and yearnings. Christopher's writing brings clarity of thought together with a spacious and compassionate view, all in the service of cultural healing."

—MARY GOMES, PH.D., coeditor of *Ecopsychology: Restoring the Earth, Healing the Mind*

"David Christopher is a Kahlil Gibran for the 21st century. His book uses the venerable tradition of the teaching-story to guide us toward a deeper understanding of the Great Transformation that is the setting for our lives. For readers with a spiritual sensibility, *The Holy Universe* offers a cosmic perspective grounded in ecological sensibility."

—RICHARD HEINBERG, Senior Fellow, Post Carbon Institute, and author of *The Party's Over, The End of Growth,* and others

THE
HOLY
UNIVERSE

THE
HOLY
UNIVERSE

A NEW STORY
OF CREATION FOR
THE HEART,
SOUL, AND SPIRIT

DAVID CHRISTOPHER

NEW STORY PRESS
Santa Rosa, California

Copyright © 2014 by New Story Press. All Rights Reserved.
No part of this work may be reproduced, transmitted, or retransmitted in any form or by any means, electronic or mechanical, including recording, photocopying, or by an information storage or retrieval system without permission in writing from the publisher (excepting reviewers who may use brief passages in a review to be printed in a magazine, newspaper, or on the Web).

Excerpts from:

The Gift of Good Land by Wendell Berry,
Copyright © 1981, reprinted by permission of Counterpoint, Berkeley, CA.

Guns, Germs, and Steel by Jared Diamond,
Copyright © 1997, reprinted by permission of W. W. Norton & Co., New York.

"Restoring Nature, Restoring Yourself" by Francesca Lyman,
Copyright © 2003, reprinted by permission of *YES! Magazine*, Bainbridge Island, WA.

Sustainability and Morphogenesis: The Birth of a Living World by Christopher Alexander,
Copyright © 2004, reprinted by permission of the Center for Environmental Structure, Berkeley, CA.

Thinking in Systems by Donella Meadows,
Copyright © 2008, reprinted by permission of Chelsea Green, White River Junction, VT.

The Awakening Universe by Neal Rogin and Drew Dellinger,
Copyright © 2009, used by permission of Neal Rogin.

Awakening the Dreamer, Changing the Dream Symposium by The Pachamama Alliance,
Copyright © 2011, used by permission of The Pachamama Alliance, San Francisco, CA.

Library of Congress Control Number: 2013941529
Publisher's Cataloging-in-Publication (Provided by Quality Books, Inc.)
Christopher, David (David M.)
    The holy universe : a new story of creation for the
heart, soul, and spirit / David Christopher.
    p. cm.
    Includes bibliographical references.
    LCCN 2013941529
    ISBN 978-0-9859339-0-6
    1. Human evolution—Fiction. 2. Evolution (Biology)—Fiction.
3. Creation—Fiction. 4. Spiritual life—Fiction.   I. Title.
PS3603.H763H65 2013   813'.6
                QBI13-600093

*Printed in the United States of America using plant-based inks on
100% post-consumer recycled paper for the interior and 100% recycled (with 50%
post-consumer content) paper for the cover, both processed chlorine free.*

Book design by Claudíne Mansour Design, www.claudinemansour.com

ATTENTION UNIVERSITIES, COLLEGES, CHURCHES, NON-PROFITS,
AND OTHER ORGANIZATIONS:
Quantity discounts are available on bulk purchases of this book (which may be used
for educational purposes, gifts, premiums, etc.). Special books or excerpts from the book
can also be created to meet specific needs. Please contact New Story Press at
531 5th Street, Santa Rosa, CA 95401 or at www.newstorypress.com for further information.

This story of the Universe
is a young story; it is changing
as we discover and rediscover more
about ourselves,
our planet,
our galaxy,
our Universe,
and the Infinite.

So you must hold this story
with a light grasp,
much as you would gently hold a bird,
trembling in your hand:
not so tight as to harm the bird
and even willing to release it
at the right time,
and let it fly off into history.

— *THE SAGE*

CONTENTS

INTRODUCTION

Like so many people who left the faiths of their youth, as an adult I still yearned to believe in *something*. But just as the religious interpretations of the scriptures I grew up with proved deeply unsatisfying to me, so too did the cold, factual stories of modern science fail to speak to my spirit. Like the Seeker in the pages that follow, I left the story of the scriptures I was given, but I also couldn't reconcile myself with the "Big Dumb Rock" story of creation put forth by the scientific worldview that emerged out of the European Enlightenment and persists to this day.

At the same time, I struggled with being in an unjust world seemingly bent on ecological self-destruction. I craved a spiritual direction that encompassed more than self-help, a story that could help me create meaningful work for myself that made a difference in a world of mass extinctions, deep injustices, and spiritual poverty.

So I quit my corporate training job, then my flying career, to seek an outlet for my passion for humanity and for protecting the ecology of Earth that also satisfied my soul and spirit.

During the exploring and teaching I've undertaken in the decade or so since, I've heard the refrain "we need a new story" from many thinkers of our time, who were themselves raised within the worldview of what I call "Modern Mind." They write and talk about how this worldview has created a global society that—for all its technological prowess and brilliance—has become dysfunctional to the point of becoming a danger to itself and to the Web of Life. And there *are* people among them—including Thomas Berry, Sidney Liebes, James Lovelock, Elisabet Sahtouris, and Brian Swimme, as well as "big historians" such as Nancy Ellen Abrams, Cynthia Stokes Brown, Eric Chaisson, David Christian, Joel Primack, and Fred Spier—who tell of a new worldview based on the latest science of the modern world, who speak beyond the traditional story of "pure" science.

There are other writers, thinkers, and activists such as Sherman Alexie, Jeannette Armstrong, Tom Goldtooth, Winona LaDuke, Oren Lyons, John Mohawk, Melissa K. Nelson, and many, many others, who have at the same time struggled to awaken those of us caught up in the overarching Story of

Modern Mind to another story that's been there all along.

Many of the ideas of these people were brought into the *Awakening the Dreamer, Changing the Dream Symposium,* created by The Pachamama Alliance in San Francisco. I became one of many facilitators of this Symposium, and I've always loved presenting the segment that illuminates the new worldview now coming into being.

But even as I dove deeper into this new worldview, this new story, I felt something missing. I missed the poetry. I missed the cadence and the sound of the scriptures. I wanted more. I wanted a story that also *felt* like the one I grew up with.

So I took my (admittedly limited) understanding of these ideas and fashioned my own telling of the story of creation, one that feels reminiscent of the scriptures and helps me make sense of the profound crises of our times.

It felt audacious, still feels audacious. I present the resulting dialogues to you simply as an offering, most certainly *not* as a final answer, an authoritative text, or a replacement for all other stories. My own teacher cautioned me to treat this work as "notes on my journey": thoughts and ideas forming a framework from which I will continue my explorations into the mysteries of life and the cosmos, yet written in a way that feels familiar and warm and welcoming. Of course, this framework will inevitably change as I continue my life's dance, especially as I explore ways of being in and relating to the world that are well outside the norms of modernity.

Although my life now doesn't always flow more smoothly than before this story came to me, and though I still carry old wounds and concerns, I no longer have such a deep nagging sense of ennui, of alienation, of being lost in a meaningless Universe. These dialogues have helped me make sense of both the science and the spirit, have helped ground me in the cosmos.

Perhaps you will find them useful, too.

—DAVID CHRISTOPHER
Palo Alto, July 2012

THE
HOLY
UNIVERSE

PROLOGUE

Story is vital to us;
        it is central to our being.

It is through stories that we find
        and share meaning in our lives;
     it is through stories
        that we teach each other
        and our young ones
           what is
               and is not important.

Our ritualistic telling of stories
     helps us in our journey
        as human beings
   by marking our life transitions
        and deepening our connection
        with each other and the Infinite.

— THE SAGE

THE MEETING

The Seeker stepped out of the bookstore, empty-handed.

He had recently lost his job.

He had lost jobs before but seldom had trouble
      finding work, for he was a reliable worker.
    Yet not only were times hard;
      the Seeker also found his heart empty.

For his unspeaking, yearning heart
    heard faint, unintelligible words
    and struggled with fleeting feelings
      echoing within—

      fleeting feelings and faint words,
      surfacing above the din
        of traffic,
        of billboards,
        of electric noise,
        of bills
          and responsibilities,
        of having to make a living . . .

    whispering, reminding him that he had not only
      lost his job,

          he had also lost his way.

The loss of income worried him,
      raising old, anxious demons
        of financial fears,
    but his spirit was simply unable
    to find the energy to look for work
    that he had long ago lost heart for.

His wife was concerned;
    she tried to encourage him
    not to worry.

But the Seeker brooded,
    stuck in a whirlpool of thoughts,
        ruminating
    about money,
    about work,
    about something better,
    not knowing what would be better,
        not daring to dream,
    then back to ruminating about money.

Tension built among him, his wife, and his daughter;
    uncomfortable silences prolonged
        the few meals they ate together,
    and talks with his wife became strained.

Tension grew within him, too,
    churning in his gut
        as well as his mind,
    until it finally pushed him
        to accept his wife's suggestion,
        and he dialed the number of the Sage.

———————————◆———————————

They had met a few months earlier at a
party of a friend of the Seeker's wife. The
Sage, who had struck up a thoughtful
conversation with the Seeker, was friendly
enough, but looked at the Seeker with a
gentle yet focused intensity that the Seeker
was unused to and made him uncomfortable.
The two sat in chairs in a sculpted garden,
and the conversation wound its way to the
topic of the Seeker's work.

> "When I was a kid, I always liked being
> outside," said the Seeker, "and figured I'd be
> a park ranger or something. But even before
> I graduated, I worked construction in the

summers. So I *was* outside, in a way, and the money was good. Really good."

"Easy to get used to, yes?"

"Yeah. Anyway, that's how it started. Eventually they promoted me to supervisor, then my company got bought up by a corporation. Now I'm pretty much doing project management most of the time."

"A long way
    from being outside,"
    remarked the Sage.

"Oh, it's not so bad. I get out often enough, I think."

"Really."

The Seeker shrugged and looked away. After a moment he said, "Well, I should probably mingle a little bit. Couple of people my wife wanted to introduce me to before we left."

"I enjoyed talking with you,"
    said the Sage,
        as they stood and shook hands.
    "Perhaps we might continue
        this conversation
        at some point?"

"Okay," said the Seeker, politely, wondering if this would ever happen. As he made his way past others at the party, he barely noticed that his heart had cracked open a little.

The Seeker stepped out of the store, empty-handed
    and empty-hearted,
        having gone in merely to pass the time
    before meeting with the Sage.

The Sun shone from behind the building,
    casting the eaves and walkways in shadow.
    Remnants of an overcast sky chilled the air in a lively way,
        as fog gave way to blue.

But the Seeker paid little attention to these surroundings,
    feeling anxious about the meeting,
    wondering if it was such a good idea after all.

He came to the coffee shop,
    and looked at the clock as he walked in.
        He bought himself a cup of coffee,
        and pulled a section of newspaper
        out of a pile in a basket next to the milk and sugar.

He walked through a hallway to the back of the café,
    as instructed, to the back patio,
        to wait for the Sage.
    When he opened the screen door, he was surprised to find
        that the Sage had already arrived.

◆

The Sage sat at the only table in the small
patio. It was a wooden table with long, thick
slats, surrounded by three wrought-iron
chairs, black paint chipped and marred.

"Ah, good morning,"
    said the Sage, smiling broadly.
    "You, too, have arrived early."

                "Yes," said the Seeker. "I don't like to be late."
                He felt slightly foolish.

"Please, sit," said the Sage,
        motioning to one
            of the chairs.

                        The Seeker placed the newspaper and his
                        cup on the table and sat. The table wobbled
                        as he leaned on it, sloshing his coffee.

"So, tell me,"
        said the Sage,
    reaching down to put a folded
    paper napkin under one of the
            table legs,
    "I had the sense that you might
            not contact me,
                yet you have?"

                        "Well," said the Seeker, hesitantly, looking at
                        the Sage's eyes, and then away.

                        "I lost my job a few weeks back."

"Ah," said the Sage.
        "And this is distressing
    to you, yes?"

                        "It's not just the job," the Seeker said. "I . . ."
                        He stopped.

The Sage waited a moment,
        then said, "It's quite all right.
    You don't need
            to say anything
    you don't wish to say."

                        The Seeker looked at the steam rising from
                        the coffee in his cup, then down at the
                        fading calluses on his hands, and then away.
                        A surge of wanting, of searching, of not
                        knowing, strained at his throat.

"I guess . . . I'm worried . . ."

The Sage nodded,
    inviting him to continue.

"And wondering."

The Sage quietly waited
    and listened.

◆

The Seeker found himself surprised
    at how much suddenly poured forth
        from his heart,
    about the unease he felt living in the world,
    about the conflict he felt
        between a sudden loss of income
          and his heart,
        which would not allow him
    to merely continue with the life
    that was unfolding before him.

His work had become devoid of meaning to him:
    creating homes for people
        he did not know,
        would never know—
    his coworkers had become tired, their
        souls long ago having
        given up pride in their work,
    all working for a corporate power
    that cared for little else than profit.

His family felt distant from him:
    rushing to and from schools,
    lessons, and practices,
        endlessly preparing,
        seldom enjoying . . .
    drowning in a sea of comforts

and electronic distractions,
yet thirsting for connection with one another.

His neighborhood felt sterile and lifeless to him:
castles having become fortresses,
strangers for neighbors
retreating behind locked doors,
shades pulled,
security signs stabbed
into pretty flowerbeds.

And the world felt anxious to him:
billboards proclaiming cures
for incessant inadequacies,
downtowns dying out,
bizarrely beautiful people broadcasting
grisly stories,
proclaiming a planet that was
dangerous,
damaged,
and not a little bit deranged.

The Seeker was bewildered;
he couldn't quite figure out
where the world went wrong,
where he went wrong,
why he felt so estranged.

This unease had crept
throughout his soul,
and now he found himself
recounting it all to the Sage,
who listened intently.

◆

"That's the big thing, really. The world just
seems so . . . messed up. I really don't know
what would make me happy." He shook his

head. "I don't know . . . I guess I wanted a
sounding board, which is why I called." He
looked at the Sage. "I hope you don't mind."

"Not at all,"
    said the Sage.
    "When I suggested we talk again,
        I meant it."

The Sage thought for a moment.

"This feeling you have,
      what was it, 'disconnected' . . .
have you always felt this way?"

The Seeker sighed. He stared at his coffee
cup, took a sip from it, and set it down,
pondering.

"Well, it's more like being out of place, I
guess. It's like somehow I don't really fit
in. Or maybe there's something about our
society, like we just don't get things right,
but I don't know what to do about it."

He paused again, recalling. "Like my
graduation ceremony. I remember how
much I looked forward to it, thinking I
would finally become an adult, that after
this magical day I would understand
everything and somehow join in the world."

The Sage nodded.
    "Yet instead you felt
        this same sense of emptiness,
    that after a ceremony that was over
            much too soon,
    and after everyone
        quickly dispersed,
    a question filled your heart,

something like 'Is that it?
    Is this all there is?'"

          "Yes, yes, exactly," said the Seeker, leaning
          forward a little. "It was so sudden, how it
          all ended and everyone left, but it felt as if
          something important was missing."

          The Seeker thought for a moment,
          remembering the ceremony. He sat back,
          took another sip from his cup, and sighed.

"Well," said the Sage,
  "there *was* something missing.
As with many ceremonies
    in our culture,
      it lacked the telling of a story,
        in this case, *your* story:
a chance to tell
    what happened to you
      during your time in school,
    what you learned,
    what you were grateful for,
    what you were looking ahead to,
    and, finally, a chance
      to thank those around you
      for their help,
      as you continued your journey.
We don't realize
  how much we need
  to tell our story."

          The Seeker shook his head. "I don't see how
          telling my story to anyone would have made
          such a big difference."

"Do you have a friend
    or a loved one
  who is about to have a birthday?

Someone who wouldn't mind
    being the center of attention?"
        asked the Sage.

                The Seeker blinked, puzzled at the sudden
                change of subject. He thought for a
                moment. "Well, yes, actually . . . why?"

"Try this.
    Organize a small party for this person.
    It doesn't need to be a complicated affair;
    the smaller and simpler, the better.

"During this party,
    after you've served the food
    but before lighting the candles
        and serving the cake,
    have the celebrant answer
    the following three questions
    to those gathered:

        What is one important thing
        that you learned this past year?

        What are two things
        that you look forward to in this next year?

        And finally, what are three things
        you are grateful for?

"Then, ask three of those gathered there
    to tell a small story, a couple of minutes long,
    about the celebrant,
        a story that illustrates something
        they appreciate about him or her.
    You might have each light a candle
        after their story.

"Then light the rest of the candles, sing the song,
    have the celebrant blow the candles out,
        and begin serving the cake.

"This should not take long, no more
        than fifteen minutes or so.

"Would you do this?"

                        "Well, I could, I suppose," said the Seeker.

The Sage laughed.
        "You don't *have* to do this, but
                a *little* more commitment
        might be in order," she said lightly.

                        "Well, yes, okay, yes . . . I'll do it," the Seeker
                        said, a little sheepishly. "But what for?"

"You shall see,"
        said the Sage, rising from her chair.
        "Trust me, we will have much to talk about
                the next time we meet.
                        Until then."

With that, the Sage touched the Seeker on
the shoulder with a smile and left through a
gate in the fence surrounding the patio.

                        The Seeker watched after the Sage with a
                        small frown.

                        "What the hell does this have to do with
                        anything?" he muttered, after she closed the
                        gate and walked away.

THE BIRTHDAY STORY

"She said we're supposed to tell stories after he answers the questions, and then have the cake, right?" asked the Seeker's wife. She was arranging celery and carrot sticks on a serving dish.

"Yep, and that's about it. Like she said, nice and simple," said the Seeker, as he finished setting out the glasses. He looked over at his wife. "So what made you think she'd be a good person for me to talk to?"

"I don't know—I don't really know her that well. I know she's not really a counselor, but Jordan and Jennifer both said good things about her, said she has a good perspective on things. Just a gut feeling I had, I guess. It seemed worth checking out."

"I gotta say, I don't think I've ever met anyone like her."

"Well, sweetheart, as strange as it sounds, I like this idea. It sounds like fun. We haven't had much fun lately."

The Seeker stopped, narrowing his eyes at his wife, considering. They looked at one another in silence. Then his face softened. "No, we haven't, have we? I'm sorry."

She smiled at him. The doorbell rang, and guests began to arrive.

Ricardo stood at the table that held his
birthday cake, his dark eyebrows furrowed
into a frown.

"Something that I've *learned* in the past year?"
he asked.

"Yes," said the Seeker, trying to sound
confident. After a few moments, he added,
"It doesn't have to be a big thing."

Ricardo shifted uncomfortably. "I'm not
sure what to say."

A sense of unease crept around the dozen or
so people in the room, who had gathered for
the party.

"That's okay," said the Seeker, nonchalantly.
"How about this: what are two things you're
looking forward to this next year?"

"Two things." Ricardo thought for a moment.
"Well, I'm certainly looking forward to
getting back on track and finishing up my
electrician certification program, and getting
past the test. And, well," he looked around
the room. "Marcia and I, we decided to
take the plunge, and are starting our own
business."

Murmurs of approval rippled through the
gathering.

"Great," said the Seeker. "Now, the last thing:
what are three things you're thankful for?"

"That's cheesy," said Ricardo. Two or three
people laughed.

"Well, I'm thankful for Marcia, of course."

"Of course," said Marcia. A few more people laughed.

"And . . ." he paused. "When Samantha got so sick last December . . ." he stopped again, cast his eyes down, and the room grew quiet.

He looked back up at the gathering. "It was such a hard time for us, and you guys were so helpful, cooking meals and covering for me and Marcia. I don't know how we would've handled it all.

"It really makes you stop when you think that your kid . . ." he paused again, and swallowed. Marcia put an arm around him, her eyes welling up.

"So I guess . . ." He shook his head. "Listen to me, *'I guess.'* No, I'm *really* grateful for all of you, and I'm so . . . God, I'm so grateful that Sammie's okay."

The gathering was quiet, reflecting.

Ricardo looked back at the Seeker. "Are we done yet?" he said, in a mock-irritated voice. The gathering burst into laughter.

"Not quite . . . now it's our turn," the Seeker said. He looked at the people in the room. "I want to ask three of you to tell a story about Ricardo. Something that shows what you like about him."

The gathering fell quiet again. People shifted in the silence.

"Okay, I'll go," said Scott, a short man with a crew cut, and the tension eased. He turned to Ricardo and said, "I just want to say thanks to you, too. When I went through that breakup a few years ago and had to move, you really helped me out. So here's a toast to Ricardo."

A round of "Hear, hear!" went through the gathering, and bottles and glasses clinked.

"I have a story," said Marcia, looking at the Seeker, who nodded.

"It was when we'd first started dating, and we took this trip to the desert to see the spring wildflowers. My parents were really uptight about it, but Ricardo was quite honorable."

A few people laughed.

Marcia continued. "Well, I wasn't the most experienced of hikers, and I slipped and fell, and really wrenched my ankle. It was late in the day, and we had a long ways to go, and I couldn't keep going. God, that hurt so bad.

"I told him to just go get help, but he absolutely refused to leave me. So we spent the night out in the desert, and it was *so cold!*"

"I was really dumb," said Ricardo. "I should have been a *lot* better prepared."

Marcia continued. "Well, anyway, he wrapped around me as best he could, kept me from freezing—thank God there happened to be a wool cap in his jacket— and the next day he helped me limp back to the car.

"For some reason, I remember the jaggedy brown rocks and the little plants poking through this desolate landscape. I felt like

we'd never make it. But he never got mad,
never complained, even though I was
probably not very pleasant."

Marcia looked at Ricardo. "So I guess . . .
what I love about him is how he's so caring
and patient. Usually."

Several people laughed amidst a smattering
of applause.

> The Seeker smiled. "One more story?" he
> asked. Three hands went up.

————————◆————————

> "Then after the *fourth* story, people still
> wanted to talk, so I just had them turn to
> each other and tell another person their
> favorite Ricardo story," said the Seeker. "We
> almost forgot about lighting the candles—
> we did that later—but it seemed like people
> really wanted to talk."

"I'm not surprised,"
     said the Sage,
      "and it's good that you gave them
           that chance to talk."

They had left the coffee shop at the Sage's
suggestion. They walked along a street in a
part of the town with older Victorian homes.

"Tell me, how long did this part
     of the celebration last?"

> "Huh . . . hard to say," said the Seeker. "It
> didn't seem very long, maybe twenty
> minutes, and I remember thinking it went
> by so quickly."

"You lost track of time, then?"

        "Yes . . . yes, you could say that."

"What does that tell you
    about adding this sort
    of story to a celebration?"

        "That it doesn't need to be very long."

"Yes, and it should *not* be very long.
    Because we are unused to such celebrations,
        even starving for them,
    we find the moment drawing us in,
    and suddenly more
        and more people want to join in,
    in spite of any fears they might have
    of speaking in front of a group.

"That is why I suggested the
    structure for the celebration:
        three people,
        two minutes,
        one story each.
    Really, these numbers
        are less important
    than making sure that you *have*
        a structure that serves the celebration,
    to help it end before everyone
        becomes aware of time again,
        with people looking at watches
        and jangling car keys."

        The Seeker nodded. "Yeah, it felt pretty good. And afterward Ricardo and Marcia told me that it was a great party. Others asked where I got the idea. I told them a friend suggested it."

The Sage nodded, smiling.
    "What else did you learn
    from this experience?"

                    "Well." He thought for a moment. "That
                    saying just a few things made the party
                    more satisfying."

"Yes,"
    said the Sage.
        "What else?
    Think back and remember.
    What happened
    as each person spoke?"

                    The Seeker thought for a moment,
                    remembering.

                    Then he looked up. "Some of the speakers
                    didn't really tell stories; they talked about
                    how much they thought of Ricardo or
                    thanked him for something. But the ones
                    who *did* tell stories, about some kind of
                    an experience with him . . . there was just
                    something better about it; everyone else
                    seemed to listen more closely."

"Good, yes,"
        said the Sage.
    "That's the most important idea,
        how compelling stories are."
    She looked at him
        and raised an eyebrow.
    "It sounds like this was
    an excellent experience."

She fell silent as they walked in and out of
the shade of the trees along the street. They
sidestepped places where the roots had
grown and buckled the sidewalk.

After a while, the Seeker cleared his throat.
"So . . . why *did* you have me do this exercise
in the first place?" he said.

"It was in response to the
    statement you made about story,"
        said the Sage.

"Oh?" said the Seeker.

"Do you recall that statement?"

"Well . . . no."

"Quite all right. You said
    that you couldn't see
    how telling your story
        would make
    such a big difference
    to a celebration."

"Oh. Yeah. Okay."

"And you learned that in fact
    it *did* make a difference,
    if you compare this celebration
    with other birthday parties
        you've been to."

"Yeah, it sure did make a difference."

"You see, we love and crave story.
    We are story-making beings.
    Ever since we learned to speak,
        thousands of generations ago,
            we have created
        and are drawn to stories.

"The reason your graduation seemed
    empty was because it *was* empty;
        it was empty of story,

and because it held no stories,
it lacked meaning.

"Think back upon your graduation,
and your feelings
when you were alone afterward.
How would it have been
had you instead gathered friends
and family together and said:

Here are some stories
about things that I have learned,
and what I have learned to do.

Here is what I have been grateful for.

Here are some people I want to thank
for helping me.

Here are some thoughts
on what I might do next
with my life's dance.

"How might your graduation
have been, then?"

"Well, I don't know," said the Seeker.
"Maybe it would work . . . but I gotta say, it
sounds hokey."

The Sage laughed.
"Of course," she said.

"This type of storytelling,
where one reveals even a small
bit of truth about oneself,
sounds hokey
because we are not used
to telling stories nor hearing others tell theirs.
We have become afraid
of telling our deeper stories,

because in their telling
    we expose ourselves, risking ridicule.
We have become uncomfortable with,
    even afraid of, the deep feelings
that sometimes accompany story."

        "That's sort of what happened at the party,"
        said the Seeker. "It felt awkward every now
        and then."

"I'm not surprised. But notice
    how we shout 'Speech! Speech!'
as we honor cherished ones
    at a celebration.
Notice how their words seldom satisfy
    when they *do* say something,
    because they don't know
        what to say,
    and we don't know
what we *want* them to say.
They say the wrong things;
    they don't say enough;
    they say too much.

"Yet something pulls us—deep down,
    we *know* that something important
        must be said;
    that is why we call for the speech.
We would do better if we called out
    'Story! Story!' and asked to hear
a small bit of the story behind the celebration:
    What happened, and what did you learn?
    What do you look forward to?
    What are you grateful for?
    Whom do you wish to thank?
And just as details
    make for good stories,

good stories
    make for good celebrations.

"Our ritualistic telling of stories
    helps us in our journey
        as human beings
    by marking our life transitions
        and deepening our connection
            with each other and the Infinite."

The Sage and Seeker continued their walk
in silence.

Presently, the Sage said,
    "Let me guess. Your next question
    might be, 'What in the world
    does this have to do with
    my current predicament?'"

                    The Seeker laughed. "Yeah, I was wondering
                    about that."

They had come full circle, stopping in front
of the coffee shop. A few passersby walked
to and fro.

"As you've seen through this little exercise,
    story is vital to us;
        in fact, it is central to our being.
    In a very real sense
        we *are* the stories we tell ourselves—
        a person cannot exist
            without a story.
    It is through stories that we find
        and share meaning in our lives;
    it is through stories
        that we teach each other
        and our young ones
            what is
                and is not important.

"We tell stories to ourselves, day in and day out,
about who we are,
what is happening to us, and why.
I'll wager that it was *this* kind of story
that prompted you to call me
in the first place."

"I'm not sure I follow you," said the Seeker.

"That's okay. For now,
simply start paying attention
to the stories around you;
we'll discuss what you notice when
we next meet."

"Pay attention to stories . . . okay," said
the Seeker.

PART I

A NEW STORY OF CREATION

The way forward is not to go back;
the way forward is
toward an emergence out of the two Minds,
a powerful union
of Ancient Mind with Modern Mind,
toward the emergence of a third Mind . . .

— *THE SAGE*

THE
GOSPEL OF THE
UNIVERSE

"But I don't *like* her," said the Seeker's daughter. "She makes me feel stupid if I don't play perfectly and I don't want to go anymore!"

"We had an agreement," said the Seeker, trying to stay calm, hands tight on the steering wheel. "We agreed that if you take violin lessons, you can go to camp this summer."

"I didn't know she would be such a *pain*."

"Sometimes you have to buck up to follow through on your commitments, even when you find out that some of them aren't much fun."

A sullen silence fell between them. The Seeker grimaced, asking himself, *What's the story here? What's the story here?*

"Tell me, then, what *is* the story behind your desire for her to learn music?" asked the Sage.

They had once again left the coffee shop and
now sat on a wooden bench toward the edge
of a large park, near a small grove of trees
beside an open playing field. Even though
it was a little past midday, the Sun had only
just started breaking through the morning
fog. The sounds of families at picnic tables
drifted from across the field.

"Part of it is that I didn't get a chance to
learn to play when I was young, and I regret
that," said the Seeker. "Another part is that
we hope it'll help her get into a good college,
maybe make her application stronger."

"And if she doesn't get
        into a good college,
she won't be able to get a good job,
and will have to settle
for a mediocre career,
        probably condemning her
to a life of misery, yes?"

"I know, I know . . . I see how *that* part is a
little crazy."

"Oh, yes—*any* story you come up with
        about the future is a little crazy.
We'll touch on this idea later,
        but for now, recognize that
        it simply doesn't serve to latch on
        to stories about the future,
        because we really don't know
                what's going to happen,
        and we end up not paying attention
                to right *here* and *now*."

"But I . . . I would feel sorry if she stopped
taking lessons."

"Goodness, yes!"
    exclaimed the Sage.
    "Music is important to the soul—
        as a child, I learned to play and sing;
        everyone in my family did.
    It was simply natural.
    I also learned how critical music was
        to my grandfather's ancestors,
        how it helped them as they struggled
            through slavery and repression.

"Ah, it is such a tragedy
    that so many in this culture
    do not dance and make music.
        It is as if part of them
        has been silenced.

"But as to your daughter,
    it might be useful
    to think more broadly,
        to figure out with her
        how music can be *fun*.
    Certainly you don't want to
        force this on her,
    but can you renegotiate
        your agreement
        with some fun in mind,
    to the satisfaction of you both?"

                    "There *is* another teacher, but he's much
                    further away," mused the Seeker.

"Or perhaps another instrument,
    or perhaps singing.
        Perhaps the second teacher might
    be willing to travel.
        Or perhaps you could help her
    have a conversation with her current teacher.
        You have options."

"Okay, I see that." The Seeker nodded.

"But let's examine one of your stories:

If I don't give my child a leg up on society,
then she will be at a competitive disadvantage,
have less opportunity, and will not thrive.

Is that so?"

"Pretty much . . . especially with the schools getting to be so bad."

"I'll leave *that* story alone for now.
This story of 'getting a leg up'—
while there is a small bit of truth to it—
causes many families
to become consumed with directing
their children's experiences,
focusing on how each experience
might help the child compete.

"But for all you know, she might decide
on a college that's not
difficult to get into.
Or perhaps she'll decide
not to go to school at all,
to follow a calling
none of you have dreamed of.
Nonetheless, your family has become
caught up in this story, turning
childhood exploration
into an adult-like competition."

"Well . . ." The Seeker crossed his arms, frowning. He thought for a moment, then his face softened and he sighed. "Yeah, I guess we do that a lot with her."

"This is one example of how
      our small personal stories
      reflect the stories of our societies.

"And surrounding *these* stories
      is the overarching Story,
            with a capital *S*,
      that people of a culture
            tell each other and themselves
            about what they deeply believe
      about the Universe
            and the world around them—
                  where it came from,
            how people came to be,
      and humanity's place in the world.

"This story is in fact so deep
      that we seldom see it for what it is.
      We unquestioningly call it 'Truth.'
      It is very real in your psyche
            and in the psyche of all
            who live within the culture
                  that created it.

"Tell me one part of this larger story,
      using your story
      about children
      and competition."

                        The Seeker uncrossed his arms and leaned
                        forward, resting his elbows on his knees.
                        "Children and competition . . . okay, how's
                        this: Life is a big competition, where there
                        are winners and losers. Dog eat dog, Darwin
                        and that sort of thing."

"Well, Darwin didn't *quite*
      say that; his words
      have been twisted over time,

but that is
   indeed our story:
  you have to struggle
and strive to be on top, yes?"

                    "You sure better, or someone else will take
                    you down."

"Now, what do we tell ourselves
  about where humanity comes from—
    our Creation story?"

                    "That depends upon your belief system,
                    doesn't it?"

"Give it a try."

                    The Seeker thought for a moment. "Well,
                    some religions say that God created us, and
                    we were banished from paradise after we
                    disobeyed Him. But then science came
                    along and said we evolved through a series
                    of random accidents."

"Close enough.
  So here you have two stories:
  one about life being created
    and ruled by an angry God
  and another where life is a random
    and ultimately meaningless,
    competitive struggle.
  Both rather bleak, yes?"

                    The Seeker pondered. "You know, when I
                    was a kid I never did really buy into the God
                    story they fed us when we went to church,
                    but I can't believe it's all just an accident."

"You threw out this story
  about the scriptures that was given to you,
    yet you can't reconcile with the

'Big Dumb Rock' story, either,
so that leaves you where?"

    The Seeker gazed skyward, past the
    trees on the other side of the field,
    thinking. Suddenly he looked at the Sage,
    thunderstruck.

    "I don't *have* a story!" he exclaimed.

The Sage laughed.
    "Well, you *do* have a story, but
    'There's got to be a better story than this'
        is hardly a foundation
            for a fulfilling life."

        The Seeker laughed ruefully.

"You see, when the sciences
    of Modern Mind really got going,
    they dismissed the religious,
    mythical, and spiritual stories
        of the ages.
    They tried to replace stories
        filled with metaphor and images
        with stories of mere facts and logic.

"But they failed, for facts alone
    don't go far in feeding
    the part of the soul
        that desires connection,
        loves imagery,
        craves meaning,
        demands metaphor.

"I'll wager *this* is why
        you feel ill at ease with the world,
    and why a new job,
    even a 'meaningful' one,
    would satisfy you for only a short time.

"The sciences of Modern Mind
    threw out the old stories
        but didn't replace them
        with truly meaningful ones;
    this lack puts you at odds
        with your spirit because you *feel* this,
    you feel the void in the middle of
    the overarching Story of Modern Mind."

            "Wow. Yeah. I can see that now," said the
            Seeker.

"There *is* another story.
    It draws ideas
        from newer discoveries of science
            that have yet to permeate our culture,
        from discoveries that go well beyond
            the old scientific stories.
    It embraces the mysterious
            and the mystical,
        just like stories of other faiths
    that humanity has created through the ages.

"It at once challenges
    the story of our modern world,
        or what I've been calling
        'Modern Mind,'
    and shows us a possible path forward.

"Perhaps you would like to hear this story;
    it might help you to reframe
    your current predicament."

            The Seeker settled back on the bench. "Yes, I
            would," he said.

◆

"This story of the Universe,"
    said the Sage,

"is a young story; it is changing
    as we discover and rediscover more
       about ourselves,
       our planet,
       our galaxy,
       our Universe,
       and the Infinite.

"Does this make sense to you?"

                "I think so," said the Seeker. "You're saying
                this story changes as we learn more and
                more."

"Yes, and this story
       acknowledges that fact.
    While we may be confident
    of many of the details,
       details will change over time
          as we learn more
          about the Infinite.
       Each era of Modern Mind
finds the 'absolute truths' of a previous era
       antiquated and quaint,
          not realizing that its own absolute truth
          will also seem antiquated and quaint
             in the next era.
This story is not the whole story,
       can *never* be the whole story.

"So you must hold it
    with a light grasp,
       much as you would gently hold a bird,
    trembling in your hand:
       not so tight as to harm the bird
          and even willing to release it
          at the right time
             and let it fly off into history."

She drew in a breath, let it out,
      and looked at the Seeker with
      a bright smile on her face.
            "Shall we proceed?"

                              "Yes," said the Seeker, leaning forward.

The Sage began.

THE INFINITE AND THE
CREATION OF THE COSMOS

1.

It was the Beginning of all Beginnings.

> It was a time of no time,
>> for time did not exist,

> a place of no place,
>> for space did not exist.

> It was a beginning like no other,
>> for there was no "before."

There was no time, there was no place;

> there was only
>> the Infinite,
>>> and the Unfathomable Mystery.

2.

And the Unfathomable Mystery was all that was to be.
> And all that was to be was small,
>> smaller than the tiniest particle,
>>> in the tiniest atom
>>>> in the tiniest
>>>>> dream,

>> although
> there was nowhere from which to see
>> how small all that was to be
>>> was,

> for there *was* no outside, and no inside;
>> no here, no there,
>> no now, no then,
>> no darkness, and no light.
There was nothing

save a whisper of what might be . . .
an infinitesimal,
indescribable,
Unfathomable Mystery.

3.

And out of the Unfathomable Mystery,
        the Infinite called forth unto Itself.

        It called forth space.

        It called forth time.

        It called forth being.

And in that ecstatic instant,
    the Universe
    slammed exploding outward into existence,
    with the jubilant
        and breathtaking
                beauty,
                brightness,
                energy,
                violence,
                power,
                and glory
    of a trillion newborn stars.

4.

It was a perfect explosion.

    One trillionth of a second slower,
        and the Universe would collapse back into itself,
        unfulfilled.

    One trillionth of a second faster,
        and the Universe would spin out
            dissipating into
            nothingness,

never to create
    galaxies,
        stars,
            planets,
                life,
  nor us.

A perfect explosion.

And in the instance of this perfect explosion,
the Infinite began a brilliant and fierce cycle
    of gloriously creating,
        violently destroying,
            and gloriously creating once more . . .

As if It knew what It was doing.

THE SEVEN CREATIONS

5.

Out of the fire of time, space, and being,
    the Infinite brought forth
        the Seven Creations:

    Forces, Particles, and Energy,
     Galaxies,
      Stars,
       Stardust,
        Planets,
         Life,
          and the Web of Life.

The First Creation—
    the forces of attraction and electricity
    together with particles and energy,
        out of which everything in the Universe
        would henceforth be created—
    flashed into existence,

        roaring throughout the cosmos
    as the fabric of time, space, and being
            raged outward
        beyond the size of a galaxy
            in an instant of time.

As the Universe rushed outward,
        many of the tiny particles
    flashed together and annihilated one another,
        swirling in a fire of energy, billions of degrees hot.
    Yet some of these particles avoided oblivion
            and persisted.

6.

But all was chaos:
    these persistent particles

spun entwined within the energy of the cosmos,
inseparable from its immense fire.

Then the heat within the Universe
began to subside, ever so slowly,
as the expansion of the growing Universe
also began to subside.
Over hundreds of thousands of years,
particles and energy
began to separate from one another.

And it came to pass,
after three hundred thousand years of cooling,
that an enormous flash of energy
engulfed the entire Universe,
as the forces suddenly—
everywhere
and instantly—
slammed together
new arrangements of particles,
slamming them together
to create atoms of simple elements
and also matter and energy
that was mysteriously dark—
setting the stage for forms of creation yet to come.

Thus did particles and energy,
guided by the forces
of attraction and electricity,
complete their separation,
and the Universe settled and cooled,
expanding across immense distances.

7.

For millions of years, the cooling Universe
expanded as an enormous, undifferentiated
ocean of atoms and energy.

Then suddenly, these atoms felt the tug,
    the call of gravity,
  and began to arrange themselves,
    spinning
      into a trillion whirlpools,
    a trillion gargantuan clouds:
    spinning into galaxies—the Second Creation,
    with even more immense spaces
      between each cloud.

Within each gargantuan galactic cloud,
    billions of smaller clouds formed;
      within some of these,
  gravity drew in and collected points
    of these tiny atoms of simple elements.

These points grew into spheres,
    which grew larger and larger
      as the gravity within them grew stronger,
  drawing in ever more atoms.

And these spheres became immense.
    Atoms crushed upon atoms,
    growing hotter and hotter
    as gravity grew stronger and stronger . . .

Until the pressure became so great
    that the atoms within these points
      ignited,
    shooting bright energy
    outward against gravity,
  forming a new balance within the Universe,
    as these spheres of simple elements
      flashed
    into the fire of stars—the Third Creation.

The stars within the galaxies
    danced whirling around each other,
    transforming the galaxies
      into brilliant spirals and ellipses

of a thousand colors and shapes:
spinning,
colliding,
merging,
until the galaxies themselves
danced into clusters
of perfect balance.
And the stars within the galaxies
began a brilliant and fierce cycle
of gloriously creating,
violently destroying,
and gloriously creating once more.

8.

For billions of years, stars burned brilliantly,
lived, and died.

Some slowly faded away.

Others died exploding
with the force of a billion stars,
into great bright vibrant flashes of energy.
From within the fire of these colossal explosions
burned forth stardust—the Fourth Creation.

And stardust—made of new,
fantastic arrangements of particles,
new, beautifully complex elements—
brought forth new and astonishing
potentials for creation.

9.

Stardust from these vast explosions
tore through and around
neighboring clouds of atoms and energy,
mixing with these clouds
and collapsing them down
to create new stars.

More stardust spun
    whirling around these new stars,
        and the beautifully complex elements of stardust
            formed new combinations,
            formed simple molecules of stardust;
    these simple molecules themselves joined together,
      forming beautifully complex molecules.

And this stardust fell together to form asteroids, comets,
      and planets—the Fifth Creation:
    offering places for
    these many forms of stardust,
    these new potentials for creation,
      to take hold
        and experiment.

THE GRANDMOTHER STAR AND THE CREATION OF EARTH

10.

After many billions of years
   of creating,
     destroying,
   and creating once more,
  in one particular galaxy, one particular star,
   our Grandmother Star,
     exploded violently
     with the force of a billion stars.
And from the death of the Grandmother Star
   burned forth stardust,
     forged in the brilliant fire of her destruction,
and her stardust scattered across
   light years of space.

11.

Hundreds of thousands of years passed, as the stardust
   from the death of the Grandmother Star
     tore through and around
   neighboring clouds of elements and energy,
     spinning these collapsing clouds
       back into themselves
  into many new points of light,
     arranging themselves into new stars.

One of these new stars was our Sun.

Bands of more stardust gathered circling,
   dancing around the newly shining Sun.
  Some of the bands fell together,
     forming into tiny swirling jewels, ringed around
      this brilliant yellow diamond of light.

Thus was the birth of Earth and her siblings.

12.

A fledgling Earth gathered stardust
        from asteroids,
        comets,
        and meteors
    as they collided with her.
    Earth grew hot with these collisions
        as she grew in size.

It came to pass that an asteroid
    the size of a tiny planet
    slammed into her,
        tilting her to one side
        and tearing from her flesh
    what would become her Moon.

And the Moon thus formed
        spiraling out around Earth,
    cooling into a bright silver sphere,
        drifting into a perfect distance—
            not too far away,
            nor too close—
    and serving to keep Earth in balance
        as she orbited her Sun.

For another billion years
    Earth gathered more stardust,
        adding to her forming crust
    and creating her atmosphere,
        which became dense
        with all manner of gasses.

And Earth churned with fire
        and heaved with explosions,
        just like her siblings.
    Some of her siblings flared out,
        becoming lifeless as stones;
    others entered endless cycles of storms
        and became great gaseous clouds.

But Earth, lovely Earth—
   she swirled around her Sun
     in perfect balance:
  not too hot, nor too cold,
     not too large, nor too small,
        not too close, nor too far away
     from her parent Sun.
  She cooled, her crust riding above
     currents of magma seething below,
     and ceaseless storms filled the seas
       and separated the continents.
  Her Moon pulled
     on these newly forming seas,
     creating just the right amount of tidal flow
  surging back and forth,
     covering and uncovering the shores
     of her shifting continents
       as they moved into just the right places.

These forces joined together,
    pulling and pushing one another
      in concert, setting the stage for
        the next unpredictable
          creation.

13.

As a cauldron,
    boiling, surging, stirring,
      Earth churned stardust together:
  stardust created in the death
     of the Grandmother Star,
  stardust crushed deep within
     the flesh of Earth,
  stardust pouring forth
     from volcanoes on land
     and from dark, boiling undersea vents,
  stardust thrashed within
     the seas and oceans,

thrashed by the lightning of storms
a hundred times more powerful
    and frightening
than any storm since.

Stardust, in the form
    of beautifully simple and complex elements
    and simple and beautifully complex molecules,
        endlessly formed new combinations
    as if it knew what it was doing.

For millions of years,
    stardust played and danced and created.
    As asteroids and comets
        hurled down more stardust from the sky,
    Earth poured forth more stardust
        from her volcanoes
        and her dark, boiling undersea vents,
    and pulled starlight energy from her Sun
        as the cycle of storms carried on.

14.

And it came to pass—
    perhaps in the shallow seas
        surrounding the continents,
    perhaps far down within Earth's crust,
    perhaps deep underwater
        near the dark, boiling vents—

    that these beautifully complex molecules
            gathered together
            and brought forth something
            both startling and unforeseeable:

They arranged themselves into a miraculous
        dance,
    as if they knew what they were doing.

Stardust joined together
in the Sixth Creation,
and danced
the new dance of life.

15.

Through the dance of life
came forth the consciousness
and creativity of life:
beings
aware of their surroundings,
moving toward food
and away from harm,
and beings dividing,
bringing out of their own flesh
new beings identical to themselves.

These tiny beings of stardust
lived, fed on the flesh of Earth herself,
and cleaved new beings,
over and over again.

Over the millions of years,
one creature changed itself
ever so slightly . . .
Then another, quite suddenly . . .

Into new beings, still very tiny,
that found novel ways of releasing
the energies bound
within the stardust of Earth.

As they transformed themselves,
they inscribed their lessons
into the sacred coiled codes of life
held deep within their cells:
lessons on how to thrive
upon a glorious Earth.

They shared these coded lessons with one another,
    learning new lessons
        as they spread across Earth.
They filled the seas
    with a diversity of colors and life never before seen,
        as if they knew what they were doing.

And through these infant circles of relationships
    between the tiny beings themselves
    and between Earth and all the beings
        came forth the Web of Life,
            the Seventh Creation.

BEYOND CHANCE AND CREATORS

The Seeker shifted in his seat, his brow
furrowed. "Can I ask you something?"

"Certainly," replied the Sage.

"You've used that phrase several times, 'as if
they knew what they were doing.' What do
you mean by that?"

A breeze stirred the trees in the grove behind
them. The Sage shifted on the bench,
adjusting her shawl against the chill.

"What I mean is that
    something other than chance
        seems to be driving the Universe,
            driving it to create;
    if chance alone were to account
    for the whole of the Universe,
    then life might happen only once
    in billions upon billions of years,
        yet we find ourselves here
        after only 13.8 billion years."

"But it still *could* happen by chance, right?"

"There are those who argue that point,"
    said the Sage.

"But since the Beginning of All Beginnings,
    greater and greater order
    and complexity have emerged
    at such speed and creativity
    that I find chance alone
    a deeply unsatisfying explanation."

"You're saying that God exists, then."

"Not quite. For me,
    the images of egoic beings
    conjured by the words
        'God' or 'Creator'
    are small, shadowy echoes
    that do not even begin
        to illuminate the grandeur
        of the Infinite.

"The word 'Infinite'—
    or, even better, the phrase
    'the creative force of the Infinite'—
    better captures the incomprehensible
        depth and complexity
    of the processes that drive the creation
    of galaxies, stars, planets, Earth, and us.
    This phrase helps me to see
        and to *feel*
    beyond both stories—
        the story of a determined universe
        and the story of a random universe—
    to a story of a mysteriously
        creative Universe."

           "I'm not sure I get it . . . it seems that you're
           just substituting 'Infinite' for 'God'," said
           the Seeker.

"But they're not synonymous;
    the word 'Infinite' describes
        more of a process
        than an object."

           "What do you mean, a process?"

"The creative force
    of the Infinite is like
        the *shape* and flow of the
    water of a river,

just as it is also the water
and the riverbed itself.
But here we go . . . can we really
use words to describe
that which is ultimately
indescribable?"

"I think that kind of helped; you're saying it's
about action and not things, right?"

"Actually, it's both . . .
but our culture stresses
the things and not the processes.
So it's okay to stress the idea
of 'process,' because you're used
to thinking more about 'things.'

"The Infinite, then,
is a creative process that,
even as It causes great destruction,
also brings amazing things
and actions into being—
all so wonderful, so complex,
that It certainly *acts*
as if It knows what It is doing.
That is the ultimate and delightful mystery."

THE
EMERGENCE OF THE
WEB OF LIFE

The Sage rose from the bench and stretched.

"Let's walk for a bit as we continue,"
    she said, picking up
        the brown leather satchel
    she had brought with her.

           "Okay," replied the Seeker, also rising.

They took a path that was wide enough
for the two of them to walk side by side. It
wound deeper into the park, past the picnic
tables and into the woods.

"So, then,"
    said the Sage, as she walked,
    "the Infinite is the container with no boundaries;
    It is the unspeakable source of everything,
    the indescribable setting for
        the Unfathomable Mystery.

"In a mysterious process that neither chance
    nor a supreme being can fully explain,
        the Unfathomable Mystery
           gave rise to the Universe.
    The Universe gave rise to the galaxies.
        The galaxies gave rise to the stars.
          Our Sun gave rise to Earth.

And our Earth gave rise to life
and the Web of Life."

"By 'Web of Life,' you mean nature, right?"
said the Seeker.

"A common term, yet again,
there's a danger inherent in it.
When modern humans say
'nature,' they imply an object
separate from humanity.

"Nothing could be further
from the truth, so I use
the term 'Web of Life'
to recapture a greater reality."

"The Web of Life is bigger than nature,"
ventured the Seeker.

"More than 'bigger.'
The Web of Life is a process of
vast mysterious intelligence
that *includes* us.

"The term evokes something
that's bigger *and* greater than
the concept that Modern Mind holds
when it says 'nature'."

"So 'nature' implies an object, and 'Web of
Life' something . . . alive."

"Exactly," said the Sage, smiling.
"And we are very much part
of its aliveness."

The path narrowed as it dipped down into
a small stream. They crossed on stepping
stones to the other side.

"Let's also look at this idea of
        'giving rise to,'
    which refers to the Essence
        of Emergence and Unpredictability,
    one of six Essences of the Web of Life.
    Let's return to the story of the Web of Life,
        and how it reflects these essences."

THE ESSENCES OF THE WEB OF LIFE

1.

The Web of Life emerged from the Infinite,
  bringing forth certain essences
      of the Universe
  in new and surprising forms.

There was no foretelling
  of the creativity of Earth
      and the Infinite
  as the Web of Life became
      more beautifully complex
          with time.

2.

So, then, as the Infinite unpredictably
  brought forth forces, particles, and energy
      out of the Unfathomable Mystery,
  brought forth galaxies
      out of the forces, particles, and energy,
  brought forth stars and stardust
      out of galaxies,
  brought forth simple molecules
      out of stars and stardust,
  brought forth beautifully complex molecules
      out of these simple molecules,
  and brought forth tiny beings
      out of these beautifully complex molecules,

so then did the Web of Life
  emerge into its infancy,
      born among
      the delicate relationships
  between the tiny beings themselves
      and between Earth and all beings.

Thus the Web of Life reflects
    the Essence of Emergence and Unpredictability:

That out of small beginnings
    and endless experiments,
    new and unforeseeable wholes
        shall emerge,
    that these wholes themselves
        shall become the beginnings
        of greater, more intricate,
            and more beautifully complex
        and unpredictable wholes,
    all in a mysterious, eternal cosmic play
        of self-organizing,
            self-discovering,
                and self-creating.

3.

Over millions upon millions of years,
    these tiny beings, too small to see,
    created new forms of themselves,
        inscribing their lessons
        into their coiled codes of life.
    They changed the flesh of Earth
        even as they fed upon her,
        as they grew, lived, and divided.
    Trillions of their tiny, invisible bodies
        drifted to the bottom of the seas,
        forming great layers of minerals
            within Earth's crust.

4.

Yet catastrophe was at hand.

For as enormous as Earth was,
    food became scarce.

Feeling the stress of famine,
    the tiny beings learned to feed
        upon the wastes of neighbors
        and upon the bodies
        of those that had died.

And it came to pass
    that one of these tiny beings,
        hungry for food,
    turned on another being,
        attacked it
            and destroyed it,

      taking its victim's stardust and starlight
        for its own.

The first predator:
    sustaining itself
        by killing another—
    and life became violent toward itself.

5.

Yet even these sources of food
        became scarce.

Then, one day, one of these tiny beings,
        hungry for food
    and feeling the stress of famine,
        responded in a way
    both startling and unpredictable:

This tiny being felt the warmth of the Sun.
    It turned toward the Sun's rays
    and seized tiny sparks of starlight energy,
        using them to strip apart
            molecules of water and air—
    and thereby creating a new,
        bountiful source of food.

Using starlight in this way,
    this new being prospered.

And with this new source of sustenance
    the tiny beings all once again thrived
        as lessons were coded and shared.
    Within some
        flowed stardust from Earth,
    within others
        flowed stardust from each other,
    and now, within still others,
        flowed starlight from the Sun.

Thus the Web of Life reflects
    the Essence of Flow:

That stardust—matter—
        shall circle within
        the Web of Life,
    while starlight—energy—
        shall pass through
        the Web of Life,
    both in an unending flow
        of creation,
    bringing forth the myriad
        beings of the Web of Life.

6.

Over millions upon millions of years,
    these tiny beings, too small to see,
        created new forms of themselves,
        sharing their coded lessons as they thrived.
    They created colonies,
        each being contributing to
            and benefiting from the whole,
        each being feeding others,
            pushing toward greater cooperation,
            pushing toward greater interdependence—
    as if they knew what they were doing.

7.

Yet once again—catastrophe:

For as the beings who seized starlight
          to strip apart water and air
     became stronger and multiplied,

          they gave off oxygen,

     a vile and poisonous fire that slowly spread
          and turned the color of the sky
               an ominous, deadly blue.

For this oxygen, this poison,
     destroyed the wastes of one
          that might have been food for others,
     rendered inedible the flesh of Earth,
     even penetrated the beings,
          and killed them.

Oxygen wrought havoc across Earth.
     It destroyed swarms of the tiny beings,
          their food,
          and their colonies of life.

8.

The tiny beings panicked
     in the midst of this catastrophe.

Some that survived
     fled, burrowing into the flesh of Earth,
          and hid in the shadows;
     others devoured one another
          in desperation—
and life hung in a precarious balance.

Then, one day, one of these tiny beings
     responded in a way
          both startling and unforeseeable.

Instead of fleeing the fire of oxygen,
    this being turned squarely to face
        this dangerous poison.
    It plunged into its fire.
    It wrestled with oxygen,
        struggled with it,
        and dodged its poisonous burning.
    It finally seized and subdued oxygen
        and unlocked its power.

Victorious, this being,
    having learned the secrets
        of the fire of oxygen,
    gained twenty times the power
        of its ancestors.

Filled with its new-found power,
    this new being of fire
        thrived and multiplied,
    creating new forms of itself
        as it spread across Earth.

One of these fiery beings,
    armed with the power of oxygen,
    heard echoes of its predatory ancestors
        and went on the hunt.
        It attacked a second being,
        penetrating its flesh,
        and destroyed it—

    taking its victim's stardust and starlight energy
        for its own.

It inscribed its violent lessons within
    its coiled codes of life
    and passed these lessons on
        to others.

9.

Yet there were new lessons to learn.

For it came to pass that two of these tiny beings
        together brought forth something
        most startling and unpredictable.

The first, a violent being of fire,
        armed with the power of oxygen
        and hungering for food,
                attacked a second being,
                penetrating its flesh,
                intent on destroying it—

yet this second being did not die.

It remained alive,
        harboring and feeding the first within itself;
        the first, the would-be predator,
                set aside its violent ways
                and in turn helped the second being,
                protecting it from the fire of oxygen,
                        even using oxygen to feed them both.

Thus the two beings joined together as one,
        using oxygen,
                the waste of another,
                as sustenance for itself.
        And this new being's waste, carbon dioxide,
                became in turn sustenance for others.

Catastrophe begat creativity;
        competition begat cooperation.

The curse of oxygen
        became a blessing.
        All the tiny beings once again
                prospered and multiplied,
                filling the oceans of Earth—

and the blue of the sky
>> transformed from the color of death
>>> into the color of life.

Thus the Web of Life reflects
> the Essence of Flexibility:

That over time
> *the Web of Life shall mature,*
>> *grow stronger,*
>>> *and become more beautifully complex—*
> *creating more species,*
>> *more possibilities,*
>>> *and more connections.*

As it grows in diversity,
> *in myriad ways*
>> *of competition and cooperation,*
> *the Web of Life becomes resilient,*
>> *able to dance with changes,*
>> *overcome catastrophes,*
>> *thrive with the passing eons,*
> *and become ever more deeply creative.*

10.

Over the ages the tiny beings thrived,
> joining together, cooperating,
>> creating new colonies of life
>>> as their ancestors had done before.

Some beings took hunting to new levels,
> surrounding prey
>> and swallowing it whole
>> and alive.
Other beings experimented
> with the ways
>> of seizing starlight to strip apart water and air
>> and of using the wastes of one as food for another.
> Each species created novelties of being

within the Web of Life,
though taking many patient generations
for these novelties to take hold.

But it came to pass that the beings
created a new, intimate way of sharing
their lessons and novelties
within their species.

For two beings of one species
learned to each uncoil their codes of life,
cleave them in half,
and entwine them with the other's—
half from one,
half from another,
to form a new child being.

In this way, the species,
within one single intimate joining
of the coiled codes of life,
learned how to pass on novelties of being
that would before
have taken a thousand generations
to take hold.

And life exploded
in diversity,
complexity,
and creativity.

11.

Over the ages the tiny beings thrived,
joining together, cooperating,
creating new colonies of life
as their ancestors had done before.

As the beings cycled stardust and starlight
within the colony,
the waste of one
serving as the sustenance for another,

these colonies joined together into a new
        level of cooperation

and became beings unto themselves.

And these new, larger,
        and beautifully complex beings thrived,
                cooperating and competing,
        creating still more new beings
                as their ancestors had done before,
        and became bigger and more intricate
                than their ancestors.

12.

Millions upon millions more years passed.
        As Earth heated and cooled
                and shifted her continents,
                life ventured forth
                        exploring those continents,
                        creating new ways of being.

And the Web of Life grew larger, more intricate,
        as it created, destroyed,
                and created once more,
        more millions of years, more beings:
                tiny fungi, large fungi,
                tiny plants, large plants,
                tiny animals, large animals,
        dancing, experimenting, improvising
                with colors, shapes, sizes, and movement—

                        oozing,
                        swimming,
                        flopping,
                        crawling,
                        walking,
                        running,
                        and flying.

Each creature created its own rhythm,
   matching the rhythm of Earth
      as she slowly turned:

      with breath, in and out,
         with days, dark and light,
            with seasons, cold and warm,
      and with eons
         of vast creation and destruction.

The Web of Life shimmered with energy,
   growing larger and stronger, becoming
        more intricate,
        more diverse,
        and more ingenious.

13.

The creatures danced a thousand dances,
      of birth
         and death,
      of violence
         and peace,
      of competition
         and cooperation.

New creatures found new ways to thrive:
      niches creating new beings,
      beings creating new niches,
   as the Web of Life grew and experimented.

The beings depended upon one another,
   some for food,
      some for protection.

   Creatures at times competed and killed
      and at times created partnerships,
   as if they knew what they were doing.

Thus the Web of Life reflects
   the Essence of Relation:

That all creatures of the Web of Life
  shall be bound to one another,
  each depending upon others
  for the creation of sustenance.

Sometimes predators,
  sometimes prey,
    sometimes partners:
  all beings are joined together
    in an interdependent web
      of competition
        and cooperation,
  wordlessly communicating
    the desires of each
  into the abundant hum of the Web of Life.

14.

And the Web of Life
  brought beauty into its creations.

It wove symmetrical tapestries
    within each being.
  It splashed perfumes and colors
  across fields of flowers
    and groves of trees
  that insects might more quickly
    find their nectars.

It purified the waters
  with the music of creeks
    and waterfalls.
  It gave melody to the birds
    and calls to the animals,
  as they sang warnings and welcomes
    to each other.

It wove tastes and smells into plants.
  Some bore sweetness,
    inviting animals to partake

of their offerings of food
in exchange for the scattering of seeds;
others bore bitterness,
warning of poisons
and whispering of medicines
held within.

It created soft furs,
strong, smooth feathers and scales,
and warm nests and dens
that its children might find comfort
and thrive.

Thus the Web of Life reflects
the Essence of Beauty:

*That beauty shall be bound throughout
the Web of Life, pointing the way
toward harmony, usefulness,
and mystery.*

*In the delicate fronds of the ferns
glowing green in the forests,
beauty shall be found.*

*In the grace of the lioness
chasing down the gazelle
as they dance the circle
of birth and death,
beauty shall be found.*

*In the magnificent violence
of the thunderstorm and the erupting volcano—
and also in the tranquil morning mists
across an infant African savanna—
beauty shall be found.*

*Beauty shall be found throughout
the Web of Life,
for it is woven throughout
the Universe.*

And one day the Universe shall stand upright
        and behold itself,
    breathless in silent awe.

15.

Yet—often:

The Web of Life saw fit to visit destruction
    upon some creatures,
        as they grew unable to navigate changes
            brought forth within the Web of Life,
    lost flexibility, were overpowered,
        and vanished—
    leaving only ghostly traces of themselves
        in layers of rock.

And—sometimes:

The Infinite saw fit to visit destruction
    across the whole of Earth:

    drawing ice down from her poles—
        with many creatures
        dying from the cold . . .

    bringing sudden changes in the patterns of
        the clouds,
        the winds,
        the rains—
            with many creatures dying
            for want of food and water . . .

    casting down a mountain from the sky,
        sixty-five million years ago,
            a bright flash slamming into
                Earth—
        sending skyward a roiling plume
        of fire, smoke,
            steam, and ash
        that enveloped Earth

and smothered many creatures,
with many more dying
in the wild extremes
of hot and cold
in the times that followed . . .

Five times since the creation of the Web of Life,
such catastrophes enveloped Earth,

Five Great Extinctions,

bringing the widespread death of beings
never to be seen again.

16.

Yet after each catastrophe,
the Web of Life would not perish.

Out of the violence and destruction
of the Five Great Extinctions,
the Web of Life arose and created anew,
bringing forth new beings
new beauty, new creativity,
with more thousands of beings
and more parts to play.

The Web of Life flourished again
and again.

Thus the Web of Life reflects
the Essence of Creation and Destruction:

*That creation cannot exist
without its twin, destruction.*

*From the first callings of the Infinite,
with the Universe exploding
outward into being—
with particles created
and instantly annihilated—*

the Infinite creates,
   therefore It shall destroy;
the Infinite destroys,
   therefore It shall create.

So, too, does the Universe.
   So, too, do the galaxies, stars, and planets,
      and so, too, does the Web of Life.

For catastrophe is the spoon
   that stirs the cauldron of creativity.
   Life, too, is both destruction
      and creation:
   the opposite of death is not life,
      but birth.
   For death and birth are the primordial facets
        of the Web of Life's
      deepest expression
   of the cosmic creative
      drive of the Infinite.

Therefore, life feeds upon life;
   some shall succumb to death
      that others might be born,
   and all shall eventually succumb
      that the Web of Life
   might bring forth
      ever more complexity.

17.

For billions of years, then, the Web of Life
   has created, destroyed,
      and created once more,
        each cycle of the eras bringing forth waves
          of creativity and beauty,
      setting the stages for
        the next unpredictable
          creations.

18.

Surviving through the fire, dust, and ice
    of the Fifth Great Extinction
    were small creatures whose blood
            was warm within themselves
        and who developed deep bonds
            with their offspring
            and among one another.

Over millions of years,
    these creatures grew larger,
    stronger, more beautifully complex,
        and more intelligent.

One of these creatures
    developed hands that could grasp
        as it travelled from tree to tree.
    Its descendants ventured forth from these trees.
        They learned to walk upright, freeing their hands
            to make ever more intricate tools and weapons.
        They learned to harness fire.
        They created more intricate sounds
            with their voices
        as they called to one another
            and began to create greater nuances
                of expression.

And it came to pass, not long ago,
    perhaps only fifty or one hundred millennia,
        that one of these upright, grasping animals of fire
        saw her reflection in the waters of the river
            as she had done many times before.

And, as before, she sensed

        that she was alive,

        that Earth was alive,

        that the Universe was alive.

She was once again filled with awe
    at the power of the Infinite
        and marveled at the creations
        of Earth and the Web of Life.

She took a stick near the bank of the river.
    She drew a spiral in the soft mud,
        and gazed intently at it
            as she felt
        the creative force of the Infinite
            flowing through her
    and through everything around her.

Feelings and images rose up
    within this powerful flow.

She returned to the clan,
    telling them of these feelings and images
        that arose within her
        by the side of the river.
    And they understood her,
        for others had also felt similar feelings
            and seen similar images.

And as the clan
    embraced these mysterious feelings and images,
        as it began to know itself
        and its place in the cosmos,
    a story began to form:

        and thus began
        the Story of Ancient Mind.

CHILDREN OF THE INFINITE

The path led the Seeker and the Sage to an
opening in the woods within the park. The
clouds had disappeared, and the Sun grew
warm on them as the chill in the air faded.
They walked to a large, flat rock bathed in
sunlight at the edge of the meadow.

"Before I continue with this story,"
     said the Sage, sitting,
     "I want to ask something.
          Where in this story
          would Modern Mind tell us
     that consciousness comes into play?"

                    "Consciousness . . . meaning awareness,
                    right?" asked the Seeker, as he sat next to her.

"Yes, awareness of surroundings,
     awareness of self."

                    "When humans came into the picture, that's
                    when awareness kicks in."

"You mean that consciousness
     emerged with humans?"

                    "Well, yeah." The Seeker thought for a
                    moment. "But I have a feeling that you're
                    about to say that's not true."

The Sage laughed.
          "Very observant of you!
     Yes, this idea reveals a deep conceit
          within the story of Modern Mind,
          the idea that consciousness
     is the exclusive domain of humanity.

"Yet even microscopic creatures
     move away from danger
     and toward food.
     Isn't that being conscious,
          being aware?"

                         "Well, I guess, but it's not like they know
                         what they're *doing*," said the Seeker.

"Really?" said the Sage.
     "I'm not going to belabor the point,
     because it is well outside your experience,
          or, I should say,
     outside the stories you've been told
          by Modern Mind.

"What I ask you to consider
     is how each species might have its own
          collective consciousness,
     and that we are a young species,
          that plants, which are
          hundreds of millions
          of years older than us,
     might have learned much
          in their own evolution
               and might have much
               to teach us."

                         The Seeker shook his head. "I don't
                         know . . . that sounds a bit weird . . ."

"Allow me to get even weirder.
     Our conceits cause us
     to overlook the stories
          of other creatures,
     the story of consciousness
     and awareness of all other creatures."

A small flock of birds made its way through
the brush next to the rock. The birds flitted

from branch to branch, softly calling to one
another.

The Sage nodded toward them.

"Perhaps birds
        are not just calling
    only for evolutionary purposes.
    Perhaps they're singing
    the stories of the day,
        retelling their own histories
    to one another—and to us,
        if we care enough to listen."

The Seeker furrowed his brow and crossed
his arms. "Telling stories . . . *that* sounds far-
fetched, really."

"Is it because this is not possible,
    or because you've
        not paid attention to this possibility?
    By 'this,' I mean forms
        of consciousness beyond
            human consciousness,
        the basis of deep intuition
            which you deny,
            which Modern Mind denies,
        even though it is through
        such mysterious intuitive leaps that
    modern science has developed."

"Well, I don't know—I'm a bit skeptical."

"Your skepticism,
    which serves you well,
        is also your curse.
    It is your first, reactive response
        to the mystery.
    Your highly developed intellect,
        which also serves you well,

is yet another curse—
endlessly analyzing data,
pursuing a logical explanation,
insisting that one *must* exist
    and is the only one worth pursuing,
      even when you see no explanation.
Modern Mind has taught you
to ignore, even denigrate,
    your own intuition,
to the point where you are a stranger
    even to yourself."

          "Ouch," said the Seeker.

"I'm sorry,"
    said the Sage, softening.
    "Perhaps that was a bit much."
      She paused a moment.

"My point is that Modern Mind's
    story about consciousness,
    even about evolution, has humanity
    at the top of some pyramid.

"But we are *not* the top
    of this enormous process,
    and the story shall continue,
      for the creative powers of
      the Infinite will continue
    for billions more years,
      long after our Sun has died."

          "But that sounds like the Big Dumb Rock
            story you talked about, where we're just
            insignificant, meaningless accidents."

"No, no, no—humanity is neither
    insignificant nor meaningless.

"For this creative force of the Infinite,
       this drive to greater complexity
          and greater order,
     has taken billions and billions of years
         to create the Universe and its galaxies,
   and billions more
      to create Earth,
      the Web of Life,
      humanity—
      and to create you.
In a mysterious and profound way,
      the Infinite
         wants Earth.

      The Infinite
         wants life.

      The Infinite
         wants humanity.

      The Infinite
         wants you.

"While you are no more precious
     than any other being in the Web of Life,
     you certainly are no less precious, either.

"Along with all of the beings of the Web of Life,
     you, too, are a child of the Universe,
   created out of sacred stardust:
     a child of the Infinite,
   which is a great, mysterious process
     that somehow *wants* you.

"Yet the fledgling sciences of Modern Mind
     denied this, and we still suffer
       from this denial.
   And the resulting alienation
       from the Universe,
       the isolation from the Web of Life,

the estrangement from Earth
and the rest of humanity,

is the greatest of tragedies,

for how in the world, how in the Universe,
could you be anything other than
yet another delightful creation
of the Infinite's primordial dreams?

"Earth, life,
and you
are all precious
beyond comprehension,
called forth by the
mysterious creativity of the Infinite."

They sat silently for a while, feeling the Sun
grow warm, listening to the birds as the
flock made its way across the meadow.

"I think I'm beginning to understand," said
the Seeker.

"Good," said the Sage,
smiling.

THE
BOOK OF ANCIENT
MIND

"Just as Earth gave rise
    to the Web of Life,
    so too the Web of Life
        gave rise to *Homo sapiens*
        and what I call Ancient Mind,"
    continued the Sage.

                "By Ancient Mind, you mean the first
                humans, right?" asked the Seeker.

"Close. What I call Ancient Mind
    is that which emerged
        through our shared
    human consciousness and learning,
        as *Homo sapiens* refined the use
        of symbols and language
        and created story.

"It is this Story of Ancient Mind
    we shall turn to next."

THE CREATION OF
HUMAN CONSCIOUSNESS

1.

Ancient Mind
>was human consciousness
>>emerging—
>was the Universe
>>in the form of a human being.

Ancient Mind sensed the Infinite within itself
>and sensed the Infinite within all beings.
>It saw the consciousness of everything around it,
>>from the tiniest pebble
>>to the Sun in the sky.
>It also sensed other worlds
>>beyond the world that could be
>>>seen, touched, and felt.

2.

Ancient Mind cultivated
>a deep connection with the Infinite.

Ancient Mind felt itself connected
>to plants and animals,
>connected to rivers, mountains, oceans, and sky.
>It saw that it was inseparable from the living creation,
>>giving to, receiving from,
>>relying upon, and deeply within
>>>the living whole
>>>of the Web of Life.

Ancient Mind also felt itself connected
>to fathers, mothers,
>aunts, uncles, nephews, nieces,
>grandchildren, grandparents, ancestors,
>>and those yet to be;

it did not see individuals
as separate from the clan.

Through its new-found powers of language,
Ancient Mind spoke of the Infinite
and humanity's connection with the Web of Life.
It collected and passed along
the lessons learned on how to thrive
within the bounty of the Web of Life
and became more powerful
within the Web of Life.

Ancient Mind brought forth stories of the Infinite,
of the spirits,
of the Creator,
of the place of the peoples in the Web of Life.

And the overarching Story
that Ancient Mind told itself
was this:

Earth is alive;
everything within the Universe
is alive.
We are but one species of many
in the grand kinship of the Web of Life.
And though we are subject
to the ways and lessons of the Web of Life,
we are still an intimate part of its family,
and through the Web of Life
we are connected
to the whole of the Infinite.

3.

Yet the story that Ancient Mind told itself
was not always one of harmony and peace.

For ancient echoes of animal ancestors
insisted on claiming
and defending territory,

and Ancient Mind at times thus told itself stories
        that those humans outside a clan
were not human, but the "other."

It sometimes was at peace with the other
        but sometimes lived in fear of the other—
                and at times it saw fit to make war upon
                and kill the other.

Clans also struggled within themselves,
        struggled with power,
        with justice, with traditions.

Ancient Mind, then,
        was at times violent and fierce,
        even as it was peaceful and loving.

And while it roamed the vastness of the spiritual Universe,
        it found no need for the precise measuring
                of the spaces of the physical Universe:
                        the size of Earth
                        or the distances between the Sun and stars.

Ancient Mind instead knew without words
        of the deep, indescribable Mystery of the Infinite,
                listening and learning
                        as the land,
                        the water,
                        the sky,
                        and all the species
                                sang with Ancient Mind,
                                sang the wordless songs
                        of the Web of Life.

And, knowing the Infinite in this way,
        Ancient Mind was at home within the Web of Life:
                        violent and peaceful,
                        loving and hating,
                        creating and destroying,
                        living and dying.

4.

Ancient Mind grew.
　　It learned to use fire for more than cooking food
　　　　and warming the night, as its distant ancestors had.
　　　　It used fire also to help call forth the spirits
　　　　　　and to tell stories by its light.
　　It hunted and gathered—
　　　　sometimes pursuing prey
　　　　　　to the point of extinction,
　　　　yet never to the point
　　　　　　of harming the whole of the Web of Life.
　　It painted the walls of cliffs and caves,
　　　　saw gods and spirits everywhere
　　　　　　and in everything.
　　It made war and made peace,
　　　　was deeply loving and fiercely loyal,
　　　　　　and fought with horrifying violence.

Ancient Mind learned how to survive in times of need
　　and thrived in times of plenty.

It learned the ways of plants and animals,
　　　　helping some plants to grow
　　　　　　in the surrounding forests and savannas,
　　　　and keeping some animals close by
　　　　　　for safety and for food.

Ancient Mind cultivated these relationships
　　　　with animals and plants,
　　seeking deeper lessons—
　　with some plants helping it
　　　　to journey to worlds
　　　　beyond the physical.
　　Ancient Mind learned from these journeys,
　　　　as its members found their place,
　　　　　　their dance
　　　　　　within the Web of Life.

5.

With each passing generation,
    Ancient Mind traveled the physical world,
        migrating by foot, by canoe, by raft.

And as it settled in each new niche,
    Ancient Mind gave attention to the Web of Life
        and became intimate with
            the plants, animals, and the seasons,
      creating deep knowledge
        that it passed to each generation.
    At times, clans would come together
      in great gatherings of peace,
        trading crafts, ideas, lessons, and stories.

For tens of thousands of years
    as Earth entered ages of cold and warm,
    Ancient Mind ventured across the vastness
      of her continents,
adapting to new environments
    and changing climes,
creating countless numbers
    of clans,
    languages,
    stories of creation,
    and cultures:
        some more violent than others,
        others more peaceful—
with all paying deep attention to
    the Web of Life.

6.

And it came to pass, only ten millennia ago,
    a mere flash of time in the life of Earth,
    that an age of cold came to a close,
      warming the Web of Life,
    and bringing forth great abundance.

In places of exceptional abundance,
    the clans of Ancient Mind began to change.

7.

For these clans saw power
    in joining together and staying close to
        these places of abundance.

They deepened their knowledge
    of certain plants,
        helping them to grow in the fertile soil,
        selecting those that were stronger
            and more bountiful
        for planting in seasons to come.

They deepened their knowledge
    of certain animals,
        taming them for slaughter,
    so that the clans would not have to hunt for meat,
        which in some places had grown scarce
            from overhunting.

They created stronger shelters
    as they settled
        for longer and longer stays.

They created ingenious ways
        to store and preserve food.
    Their numbers grew
        as they gathered more of the bounty
        of the Web of Life around them.

And as these clans joined together
        and created settlements,
        reshaped their surroundings,
        traded crafts, ideas, lessons, and stories
            with clans in other settlements—
    as they brought forth and multiplied the bounty
        of the Web of Life many times over,

and passed down their lessons
        from one generation to the next—
they began to see themselves
        as separate from
            the Web of Life,
and began telling a new story
        of the special and separate place
            of human beings
            within the Universe.

Thus began the first chapter
        in the Story of Modern Mind.

OUR FOUNDATIONAL HERITAGE

The rock was warm from the Sun. As
Earth turned, some tall redwood trees
nearby began to cast a shadow.

"Ah, shade . . . how nice; I was
    beginning to get hot,"
        said the Sage.

She pulled out a water bottle from her
satchel. Ice clinked against its metal walls as
she drank from it. She then offered it to the
Seeker and removed the shawl from around
her shoulders as he drank. He handed the
bottle back to the Sage, who had dug out two
sandwiches. She handed one to the Seeker.

                "Thank you," said the Seeker, accepting the
                    sandwich.

They ate in silence, enjoying cool and warm
breezes drifting gently around them.

"So, then,"
    said the Sage,
        as she finished her meal,
    "this is our foundational heritage,
    before we began to see ourselves
    as separate from the Web of Life:
        tens of thousands of generations
    of being deeply connected with
        the Web of Life,
    and deeply aware of its lessons."

                "It sounds idyllic," said the Seeker.

"Oh, yes—and what would you say about
    a society that is constantly at war,

in which a young man
        constantly faced death
            at the hands of another,
a society with a murder rate
    higher than that found
    in many modern cities?"

                    "*That* sounds messed up. I doubt it would
                    last."

"Yet this particular society
    I speak of lives now, deep in
        the Amazon forest,
    fully immersed in Ancient Mind,
    and has flourished for centuries,
        if not millennia.
    Until it recently chose
        to make contact with Modern Mind,
            it indeed lived with such levels
                of violence.

"The ancestors of ours
        who were of Ancient Mind
    were as human as those
        of Modern Mind,
            living with both light and shadow.

"Remember, to be human
    is to embody both the compassionate
        and the merciless,
    the peaceful
        and the violent,
    to be at times full of love and loyalty,
        and to be at times full of hate
        and destruction."

                    "Wait a minute," said the Seeker. "I thought
                    you said that Ancient Mind was in harmony
                    with the world, which is what I've thought

all along. Yet you're also saying that the
peoples of Ancient Mind fought and killed
each other as much as we do now."

"We must be careful
    not to over-generalize;
    the thousands of cultures
    of Ancient Mind are as varied
    as the thousands of flowers
        in a rainforest.
    Not all cultures within the depths
        of Ancient Mind have embodied
        such levels of violence
        as a way of life.

"Indeed, another culture
    of Ancient Mind, the Haudenosaunee,
    or Iroquois, after generations of violence,
    consciously created a strong social structure
        to harness the passions
        of violence and anger.
    Their society became committed to peace,
        which has lasted many generations.
    It was so profoundly egalitarian
        that those men who created
            the founding documents
            of the United States
        saw fit to emulate much of the law
        created by the Haudenosaunee.

"In many other instances,
    while Ancient Mind
    banded together in clans,
        with deep loyalty to them,
        it was also sometimes deeply violent
            to those outside the clan—
    and strife within the clan most certainly
        was not unknown."

"So why do we still find the idea of Ancient
Mind attractive?"

"Despite the struggles and acrimony
        within the clan
    or between one clan and another,
    Ancient Mind tended to be
    deeply in tune with the Infinite,
        indeed *had* to be more in tune
        with the Infinite,
    if it was to survive and thrive
    within the Web of Life.

"And being in tune with the Infinite
    gives one a profound sense of
        one's place in the Web of Life,
    which is missing in the hearts
    of many in the societies
        of Modern Mind."

The Seeker nodded vigorously. "That's
it exactly. I don't really feel any sense of
connection with the world, and when I hear
you talk of people being part of a larger clan,
with aunts, uncles, cousins . . . it reminds
me of how alone I feel, sometimes."

"Fortunately, while it is
    challenging for you
    to feel this deeper connection,
        it's not beyond your grasp.

"For now, let's focus
    on the emergence of Modern Mind
        out of the thousands
    of cultures of Ancient Mind."

THE
BOOK OF MODERN
MIND

"As we shall see," said the Sage,
    "the word 'separation' figures greatly
    into the story of Modern Mind.
    What thoughts does this word
    bring to mind?"

                  "Let's see . . . that something lies apart from
                  something else, that they're not connected,"
                  replied the Seeker.

"Yes, and it is important
    to point out that *separation*
    is not quite the same
    as *differentiation*.
One can be connected
    to something and still
    be differentiated from it.
You can see this even from
    the creation of the galaxies,
    as stardust gathered here and there,
    differentiating itself from other clusters;
        yes, they are different,
        but they are still connected
        to one another
    through the force of gravity.
Even Ancient Mind, as it became
    aware of itself, began to

see individuals as differentiated
from one another but still
connected with each other
and with the Web of Life."

> The Seeker picked up a short, thin twig, and twisted it as he thought. "Okay, I think I get it: Ancient Mind saw things both ways; differentiated but still connected. Modern Mind sees things as separate and *not* connected."

"That's a good way of putting it.
    It is this sense of separation,
        of differentiation
    to the point of disconnection,
    that is both the brilliance
        and limitation
    of Modern Mind.
        Let's pick up at the point where
    the Story of Modern Mind emerges."

A STORY OF SEPARATION

1.

Modern Mind created new ways
   of being in the world,
  ways different
     from those of Ancient Mind.

It settled in villages, leaving behind
   the wandering ways of Ancient Mind,
  and exploited the abundance
     of the lands surrounding it.
  More food brought forth more people,
    more people, more food—
    and populations began to grow.

Modern Mind expanded, cutting and burning down
   the groves of nearby forests—
  which were sources of food
  for all the beings in the Web of Life—
  and created larger fields,
    which were sources of food
      only for itself.
  It turned the waters of creeks and rivers
    toward these fields.

It bred plants and animals to its desires.

Modern Mind judged some creatures—
    insects, burrowing animals, wolves, coyotes—
as not worthy of their place within the Web of Life.
   It called them pests
     and sought to destroy them,
    that it might keep
    the bounty of the Web of Life
    for itself alone—
   forgetting the ways in which

these animals and plants
belonged in the Web of Life.

Modern Mind also heard ancient echoes
of animal ancestors insisting on claiming
and defending territory,
and at times made war upon its neighbors.

2.

But Modern Mind also continued
to add to the lessons it had learned.
It created tools
to grow more food,
to build more intricate dwellings,
to use as weapons.
It learned new ways of tending
its fields and livestock
and storing great quantities of food.
It harvested ideas from the stories of Ancient Mind,
stories about the spirit worlds.
It created new stories of gods and goddesses
and built shrines to them.
It formed networks of trade in times of peace,
as it exchanged crafts, ideas, lessons,
and stories—
with each exchange
revealing more powerful lessons.
It chose leaders
to settle disputes,
to bestow justice
and surpluses of food,
to command in times of war.
It chose holy ones
to see to the appeasement
of the gods and goddesses,
that fields might flourish,
livestock might be strong,

children might be plentiful,
and the village might thrive.

Wherever the Web of Life
gave more bounty
to harness
and control—
the rich soils of the Fertile Crescent,
the warm, lush river plains flowing into China's seas,
the fruitful jungles of Polynesian islands
and Mesoamerica—
Modern Mind arose and flourished.

Modern Mind saw itself as separate from
and above the Web of Life,
believing it could challenge, subdue,
and hold dominion over it,
instead of feeling at its mercy.

And the overarching Story
that Modern Mind told itself in this first chapter
was this:

Humanity is the grandest
creature in the Universe,
and the earth was created solely for its sake.
Humanity is separate from the Web of Life,
is the ruler of the Web of Life,
and can take whatever it wishes
from the Web of Life—
without consequence.

Modern Mind began to turn a deaf ear
toward the songs of the Web of Life,
believing itself and its story
to be superior to the Web of Life,
with the bounty of the Web of Life
having value only as it served Modern Mind.

3.

Modern Mind grew,
    as villages brought forth chiefdoms,
    and growing power led
    to more complex societies.

Leadership passed from parent to child
    as rulers surrounded themselves
    with those in specialized roles:
        priests,
        warriors,
        artisans,
        and traders,
    all coordinating the many tasks
        that needed tending to—
    while the multitudes
        were cast into labor upon the land.

Priests formed religions out of the stories
    of the gods and goddesses,
    and raised up grand temples
    to gain their favor
        and to sanctify the power
        of those who ruled.

Modern Mind required more children
    to fulfill its growing desires,
    and the burden of childrearing
        fell more and more upon women,
        their power diminishing
        as their domain grew smaller,
    with men slowly garnering greater power.

Ideas and lessons flowed
    from one chiefdom to another
    as Modern Mind spread,
        learning as it grew,
            growing as it learned.

It took milk and wool from some animals;
    it enslaved other animals to turn the soil
        and carry heavy burdens,
    allowing it to expand into new lands
    and push out those peoples
        of Ancient Mind that it found
        upon those lands.

And as Modern Mind collected
    ever-increasing knowledge of
    how to command more of the bounty
        of the Web of Life,
    it passed these lessons on
        to the ever-growing generations
        that followed.

4.

Yet as Modern Mind grew more powerful
    and its societies more complex,
    it became vulnerable,
    at the mercy
        of changes within
        and changes visited upon
        the Web of Life.

For in some places,
    as the rains disappeared
        and the rivers ran low,
    Modern Mind faced famine,
        having long forgotten the flexible ways
        of Ancient Mind.

In other places, as Modern Mind harmed
    the Web of Life,
        it harmed itself.
    The soil, exhausted,
        no longer grew food.
    The waters, fouled and brackish,
        no longer slaked thirst.

The landscape, stripped of trees, plants, and animals,
could no longer offer its bounty
such that shelter and tools
could no longer be made,
nor food gathered and grown.

And in those places, populations collapsed
and Modern Mind disappeared,
leaving enigmatic statues facing the seas,
great pyramids cloaked within jungles,
and ghostly dwellings rising into cliffs.

But in other places, Modern Mind flourished,
reveling, glorifying in itself
and its story.

5.

Modern Mind grew,
as chiefdoms brought forth states and empires.
Rulers used ever-increasing power
to coerce, exact tributes,
and assemble armaments and soldiers.

In places where it could travel quickly
across great distances,
Modern Mind brought forth war and slaughter;
the strong conquered the weak
and increased wealth
through the plunder of others.

Modern Mind turned tons of soil
and ripped precious metals and stones
out of the heart of Earth.
It enslaved humans and animals
on a scale fantastically beyond
any imaginings of Ancient Mind.

Yet Modern Mind also flourished in times of peace,
exchanging seeds, livestock, and ideas,
creating many new cultures.

It built cities to shelter its growing multitudes,
  creating within those cities
   immense monuments
   to rulers and gods.
It turned the courses of great rivers
  to bring ever more water to itself and its fields,
   bringing forth fantastic amounts of bounty
    from where there once was only dryness
    and drought.
It created symbols and tablets
  to account for surpluses
   and record its lessons and histories,
  tremendously magnifying
   its power to learn, recall,
   and pass on its lessons.

6.

Yet there were wanderers
  within these cultures of Modern Mind
  who heard the echoes
   of the wordless knowledge of the Infinite,
  those whose hearts felt the ache
   of disconnection
    whispering beneath
   the ways of Modern Mind.
They sought to regain this knowing,
  wandering away from the walls
  and structures of Modern Mind,
  discovering
   ancient connections
  with the Infinite and the Web of Life
deep within their spirits.

They cultivated these connections
  and taught others to do the same.

They passed these lessons on
  to others who also yearned for connection.

But even these lessons were not safe
    from the juggernaut of Modern Mind,
    as it insidiously surrounded
        and strangled lessons and knowledge,
        replacing them with its own story.

Modern Mind buried these lessons
        within religions of rules, rituals,
           and hierarchy
    and used these religions not to liberate,
        but to control the multitudes
    as Modern Mind sought power and glory.

7.

And it came to pass,
    mere centuries ago,
    an instant of time ago,
that Modern Mind began to explore
        the very large
        and the very small.

Thus began the second chapter
        of the Story of Modern Mind.

DEEPENING SEPARATION,
COURTING CATASTROPHE

8.

As it explored the very large
    and the very small,
    Modern Mind became drunk with power
    and placed its faith in reason
        and science alone,
    even splitting the two away from
        and holding them above
        all other sensibilities.

Modern Mind disdained emotion and intuition,
    rejected the old stories of its religions,
    and ridiculed the spiritual lessons
        of the plant and animal worlds,
            denying the spirit that dwelled
            within itself
            and within all.

Where Ancient Mind said
    "forests,"
    "rivers,"
    "oceans,"
    "air,"
    "Earth,"
Modern Mind said "resources,"
    and claimed them all for itself.

Where Ancient Mind saw
    a world that was alive and sacred,
    worthy of respect and reverence,
    telling a wordless story
        filled with meaning and connection,
Modern Mind saw a world for its domination:
    a world not sacred, not alive,

but to be used
as Modern Mind willed.

And when more powerful cultures of Modern Mind
        encountered weaker cultures,
    or encountered those who were still of Ancient Mind,
        it slaughtered them on a horrifying scale
            with frightening weapons and fierce diseases.

When peoples of Ancient Mind managed
        to survive these atrocities,
    Modern Mind subdued,
            enslaved,
        and attempted to assimilate them—
    and to co-opt their cultures.

Ancient Mind, pained by the madness
    of Modern Mind,
    struggled against it,
        even sought
    to enlighten Modern Mind
        of its errant ways.

But Modern Mind refused to listen,
    could not listen,
            for it *knew* its story to be beyond dispute—
            and this overarching Story
    that Modern Mind told itself in this second chapter
        was this:

The Universe is dead;
    the earth is only an insignificant rock
        spinning off into a lifeless cosmos.
    Humanity is a mere accident;
            indeed, all of life
        is a vacuous accident.
    And while the Universe somehow creates
            many fantastic things,
        and humanity is the pinnacle
        of this improbable process,

humanity must struggle—
  in unending competition—
  and must dominate the Web of Life;
  for humanity is ultimately alone in
    and alienated from
and must find its lonely stumbling way through
  a decaying and meaningless Universe.

Thus Modern Mind set out
  to conquer and exploit what it saw
    as a dead Universe
    that held no stories.

9.

Modern Mind surged across continents
  and seas.

It harnessed fire in new and powerful ways,
  first with the oils from the flesh of whales.

But as it harnessed the coals and oils
  from deep within the flesh of Earth herself,
  Modern Mind flooded across all of Earth.

It created enormous buildings filled with ingenious machines,
  machines able to do the labor of hundreds
    in fractions of time,
      creating cloth,
      lumber,
      steel,
      and astounding varieties
        of goods.
  Peasants were drawn in from farms
    as machines replaced animals
      for tilling the fields
  and as promises of affluence filled the dreams of all.

Governments encouraged the building of wealth,
  with trade giving way to commerce.
  Ideas and inventions flashed in a frenzy

across nascent networks of commerce
that extended Modern Mind's
reach around the globe.

Yet while many of its numbers came to enjoy lives
of physical comfort unimaginable even
mere centuries ago,
many more were cast into depths of poverty
never before seen on Earth.

And as it worked so triumphantly,
Modern Mind caused havoc across the Web of life,
dumping smoke into the air,
pouring poisons into the waters,
and scarring the face of Earth.

Modern Mind was blind to the harm
it inflicted upon humans and upon the Web of Life,
and, in its blindness,
convinced itself that it created good,
the world becoming better
because of it.

10.

Yet even as Modern Mind slaughtered, enslaved,
and oppressed countless millions,
there were compassionate ones
within the societies of Modern Mind
who saw and recoiled against
the stupendous horrors and deep inequities
Modern Mind wrought upon humanity.

And there were those among the oppressed
who refused to remain silent,
even at the threat
of the gun, the prison, and the gallows.

Both sought to change the ways of Modern Mind.

At first, their protestations
    fell upon deaf ears.
  Yet over the centuries, these compassionate ones
    and leaders of the oppressed
      grew in numbers
    and their voices began to be heard.

Slowly,
  painfully slowly,
    Modern Mind began to acknowledge
      the great harm it caused,
      the inequities it perpetrated,
      the deep wounding it inflicted
        throughout its history.

Slowly,
  painfully slowly,
    Modern Mind began to redress these inequities,
    and, at times, protect those
      who were previously cast aside.

11.

Modern Mind also created.

It created fantastic buildings and temples
  of glorious architecture,
  solemn, ornate, and uplifting,
    echoing the ancient knowledge
    wordlessly whispering in the spirit.

It created music of sublime complexity,
  painted paintings, carved sculptures,
    wrote plays and poetry
      to delight, entertain,
    and bend the boundaries
of Modern Mind's thought and emotion.

It created tools to see ever deeper
  into the workings of the Web of Life
    and the Universe.

And with these tools, Modern Mind learned.

It learned the workings of the human body,
    how blood flowed, carrying
        the nourishment of the Web of Life
        throughout each cell of the body,
    how catastrophic illnesses
        might be kept at bay,
    how broken bones and torn flesh
        might be better set and healed.

It learned how plants
        captured energy from the Sun;
    it learned that the Sun itself was a star
        and learned the laws of the movements
        of the planets and stars in the sky.

It learned that many more stars were in the Universe
    than human eyes alone could see.
    It discovered galaxies—many millions of them.

12.

Yet with each advance of its understanding,
    Modern Mind rejected another strand
        of intuitive knowing:

For Modern Mind saw stars, not *stars*;
    a sun, not our Sun;
        a planet, not our Mother.
    It saw a lifeless universe,
        not a Universe of consciousness,
            potential,
            and mystery.

In its denial of its connection
    with the Infinite,
    Modern Mind starved its spirit,
        creating an unsettling
            alienation within,

made manifest by those of its numbers
    who suffered from
    a deep spiritual illness and poverty
    even in the midst of physical plenty.

13.

It came to pass that the powers of Modern Mind
    threatened the very foundation
      of the Web of Life.

And the Web of Life began to collapse,
      as it had done five times before,
  with species disappearing at such rates
that had not been seen
    since the Fifth Great Extinction,
    since the mountain from the sky
      slammed into Earth
      sixty-five million years ago.

Even as voices began to sound the alarm
    that great peril was at hand,
    Modern Mind stumbled arrogantly on,
      dangerously oblivious
        of the abyss
    to which it had unintentionally
      pushed the Web of Life—
    a cataclysmic abyss
that began to engulf the Web of Life,

      thus threatening
    to annihilate humanity.

BEYOND THE STORY
OF THE FALLEN MIND

"This is where we now stand,"
    said the Sage.

        The Seeker was quiet. He still had the twig
        in his hand, twisting it as he contemplated.
        He then tossed it a few feet away from the
        rock. At length, he said, "Looks like Modern
        Mind has really messed things up."

"Looks that way, doesn't it?"
    said the Sage, lightly.

        The Seeker furrowed his brow and regarded
        her quizzically.

"Modern Mind is wrecking
    the planet—or more accurately
    wrecking the ability of
    the Web of Life to support us."

        "That's what the story says, right?"

"Yes. But I sense that you believe
    that Modern Mind is therefore
        flawed in carrying on
        in such a way."

        "Like you said, it looks that way."

"You might even say that
    Modern Mind itself is flawed,
        even fallen, yes?
    What story does that word
    'fallen' remind you of?"

        "Obviously, the story of Adam and Eve, of
        the fall of man—ah, humanity—and getting

kicked out of the Garden of Eden," replied
the Seeker.

"Yes, many who decry
the acts of Modern Mind
hold that humanity is flawed.
Look at our destructive behavior
toward the Web of Life
and toward each other;
humanity *must* be flawed.
Otherwise, why would it
wreak such havoc across the globe?"

"So isn't that a problem?"

"It is, but it's not due to any
inherent *flaw* within humanity.
We're doing what any species does:
we're following our
biological and natural imperative
to reproduce to the limits
of our habitat,
to fill our ecological niche."

"But we've overshot, haven't we?"

"Tremendously. We've cleverly used
tools and fuels to vastly exceed
the ability of the Web of Life to support us;
but again, it's not as if
there's something *wrong* with us.

"Look at it this way.
Consider your daughter
when she was an infant.
Did you think she was flawed
when she stumbled and fell
as she learned to walk?"

"Of course not," said the Seeker.

"As she grows older,
        experimenting, learning,
                even getting into trouble
                and making mistakes
                as she ventures into the world,
        are these mistakes flaws in her being?
        Is she defective?"

                                "No, but it's a problem if she hurts herself."

"Ah, but if she were never hurt,
        and you never let her risk being hurt,
                or if you sheltered her
        from all conceivable danger,
                could she develop fully
        as a human being?"

                                "When you put it that way, no, I guess she
                                couldn't." The Seeker pondered. "Wow. So
                                we *have* to let her make mistakes."

"A challenging task
        for any parent, yes?
        Now consider your daughter
        as she enters adolescence.
        What are some stereotypes of
        adolescents in our culture?"

                                "Hah, well: they're impulsive, they expect
                                instant gratification. They don't listen,
                                they're reckless and act as if they're
                                immortal."

The Sage laughed.
        "*That* came easy to you.
        What other characteristics
                describe this stage of life?"

"Let's see . . . they're concerned with
appearances, with fitting in . . . and *very*
concerned about their sexuality."

"Well, issues of sexuality
are of concern to humans
more than just through adolescence,
but as far as the deep procreative drive,
yes, this is the time
when it awakens and asserts itself."

The Sage was quiet a moment.

"Is that it? Nothing positive?"

"Oh," said the Seeker. He thought for a
moment. "Well . . . hmm."

"Interesting how the positive
aspects don't quite
spring to your mind, yes?"

"Yeah." The Seeker smiled. "Okay, teens
want to explore. They take chances, have
lots of energy. Sometimes they're very
idealistic." He thought further. "And I guess
you could say impulsivity is another word
for spontaneity."

"And all the while,
they are differentiating from their parents,
sometimes to the point of disdaining
and ridiculing them."

"Oh, *that's* certainly true," said the Seeker.

"At the same time, as they explore,
they make mistakes:
some catastrophic, some less so,
with each mistake containing
lessons within it,

if the child chooses to heed them
　　and has elders to help her
　　through the process.

"Now, adolescents in different
　　cultures behave differently;
　　if a society is dysfunctional,
　　then children will express dysfunctions
　　as they pass through adolescence.

"Nonetheless, when you consider
　　these characteristics we've outlined,
　　　　how might they apply
　　　　to Modern Mind?"

　　　　　　　The Seeker thought for a moment, and then
　　　　　　　laughed. "Wow. It almost fits us to a tee,
　　　　　　　doesn't it?"

"An interesting metaphor, yes?
　　All of these characteristics,
　　　　the ones we decry
　　　　as well as the ones we admire,
　　describe Modern Mind
　　to an astonishing degree."

　　　　　　　"Humanity's in its adolescence."

"Well, I'm not sure if that's right.
　　Yes, I've heard it said
　　that, developmentally,
　　　　humanity is in its adolescence,
　　but a species doesn't *develop*
　　　　along certain life stages
　　　　like individuals within
　　　　a species develop.
　　A species *emerges,* it *evolves.*

"However, it is possible that the *societies*
　　of a species might themselves develop.

For instance, one wonderful researcher
    has learned that colonies
    of a certain species of ants
have their own developmental patterns;
the colony grows, matures, and declines,
'living' for fifteen years or so
    until the queen dies—
even though each individual worker lives
    only a couple of years.
The age of the colony
    somehow passes on
    from one generation to the next
    in what we think are relatively
        unintelligent creatures
    with rudimentary communication—
        another example of the
        myriad intriguing mysteries
    of the Web of Life.

"At any rate, just as these colonies
    have their own course of development,
so might humanity's societies
    and worldviews,
including Modern Mind."

              "Well, for that matter, I've also heard
              that we act like adolescents because most
              adults never really grow up; we're stuck in
              adolescence because there aren't very many
              real adults around."

The Sage nodded.
    "That could very well be true, too.
        But however we cast
        the phenomenon of Modern Mind,
        these metaphors point us to
        an alternative story to the one of
        'Bad, Bad Modern Mind.'

They serve to recast Modern Mind
not as fallen or flawed,
    but instead as young,
    growing,
    learning,
    testing the limits
        of its potential,
but also rebelling against
    the Web of Life
to a dangerously dysfunctional
    point of separation."

        "Wait . . . you're not saying this justifies what
        we've done to the Web of Life, or to each
        other, are you?"

"Are an adolescent's mistakes justified,
    no matter how horrendous?"

        "Of course not. There has to be a line drawn
        somewhere," said the Seeker.

"Exactly. A balance must be struck.
    An adolescent must be given room
    to grow, to make mistakes, to learn—
        and to learn accountability
        for one's deeds and mistakes.
    Perhaps, as Modern Mind has emerged,
        the Infinite has allowed it
        room to learn and make mistakes.
    But we must also learn accountability,
        accept responsibility,
        make restitution,
        and become reconciled
    when our actions
    have caused deep harm.

"For Modern Mind is as
    a drunken adolescent,

slowly sobering,
    wanting to run and hide
as she enters the doors of the hospital
to witness the carnage
that she has wrought
    by driving her car at high speed
    into a van filled with children."

                    The Seeker blanched. "Good God . . ."

The Sage cocked her head.
    "Does this picture make you
        uncomfortable?"

                        "It does seem a bit over the top," said the
                        Seeker, raising his brow.

"Yet my meager metaphor *pales*
    in comparison to what
    Modern Mind continues
        to perpetrate,
    unintentionally or not."

                    The Seeker looked away. "I see your point."

"We must face the truth squarely:
    we have inflicted tremendous
    pain and suffering upon
        ourselves,
        our brothers and sisters,
        our fellow creatures,
        and upon Earth herself,
    but the greater tragedy
    would be if we failed to awaken,
        failed to learn from our mistakes,
        failed to accept responsibility."

                    "So it's time for us to grow up."

"Or wake up. Rather sobering,
    when you look at it this way.

The story of humanity being flawed
    or fallen not only
inflames a sense of separation
from the Web of Life,
    it also lets us off the hook,
    makes excuses for the inexcusable.
This new story challenges us to face up
    to what we've wrought."

        "Damn." The Seeker shook his head, leaned
        forward, putting his elbows on his knees,
        and stared at the ground in front of him.

"Come now, it's not all darkness
    and seriousness.
    Hidden within catastrophes
        are gifts and opportunities;
perhaps we have the chance here
        to awaken potentialities within us,
        perhaps to live
the most meaningful lives
that we could ever hope to live."

        The Seeker looked up at the Sage and
        frowned. "The most meaningful lives?
        How?"

"We'll talk about that soon,"
    answered the Sage.

"But first, let's look
    at another aspect of the story
    that says there's something
    fundamentally wrong
    with Modern Mind."

THE GOSPEL
OF PLANETARY
MIND

It began to grow cool again in the shade
where they sat. The Sage picked up her
shawl and wrapped it around her shoulders.

"The questions are these:

Why did Modern Mind attempt
to sever itself from the Web of Life
and from the Infinite?

And why did those recent ancestors of mine,
who themselves were of Ancient Mind,
sustain a deep connection to the Web of Life?
Was there something different about them?

What was different
about those of *your* recent ancestors,
who, through the European empires
of Modern Mind,
developed miracles of technology
and social structures that enriched it,
but at the same time
wrought tremendous abuse
upon others and the Web of Life?
Why did they attempt
to wipe out Ancient Mind
wherever it was encountered?"

The Seeker shifted uncomfortably.

"A rather touchy subject,
    to say the least, yes?"

                    "Well, yeah, it is. But in answer to your
                    question . . . I don't know . . . there *are*
                    people who would say there actually was
                    something different about people in
                    different places."

"Indeed, they're partially correct,
    but not for the reasons
        you might imagine.
    For the difference was *not* in
    the genetic makeup
    nor the content of the characters
        of our ancestors
    so much as the physical places
    in which they lived."

                    The Seeker frowned. "I don't understand."

"As Ancient Mind emerged
    from the Web of Life,
    it formed into tribes and clans,
        where leadership tended to be
        informal and decentralized,
        and fell upon those who proved
        themselves to be leaders.

"Modern Mind emerged out of
    Ancient Mind, but only in those
    places on Earth that *allowed*
    populations to grow and create
    more complex social structures,
        where leadership became hereditary,
        societies became stratified,
        and power became centralized.
    These cultures developed worldviews
    of separation from the Infinite."

"So they got powerful, and they also had
worldviews that justified the use of that
power," ventured the Seeker.

"Yes, and that power
    was used to subjugate
    its own people
    and the people it conquered.

"However, if the bounty of the Web of Life
    or the geography of the environment
    did *not* allow for such increases
    in population or the development
    of these complex social structures,
    and if cultural forces did not seek
        to glorify the human over all,
    then Ancient Mind remained intact,
        though its cultures
    certainly continued to evolve.

"Simply put, history followed
    different courses
    for different peoples because
    of differences in peoples'
    environments and cultures,
        not because
    of biological differences
    among peoples."

"You're saying that there wasn't anything
    inherent in your ancestors that caused them
    to be overrun by Modern Mind?"

"Just as there was nothing
    inherent in your ancestors
    that caused them to do
    the overrunning."

"Damn." The Seeker pondered this.
"That's . . . well, I don't know . . . for the

longest time I've felt a weird sort of guilt,
like there really was something almost evil
about my ancestors."

"Yet they were no more human,
       and no less human,
              than mine."

"That still feels weird, like it somehow
excuses five hundred years of genocide."

"Well, *much* more
       than five hundred years, really.
       But naming it, understanding it,
              doesn't *excuse* it—
       it merely *explains* it, at least partly.

"And there is much work
       for you to do, for us all to do,
       to acknowledge the privilege
       that you and others have
              as a result of this history,
       to come to terms with
              what Modern Mind has wrought,
       and to attempt to heal the wounds
              and dismantle the structures
              that oppress those who do not
              enjoy such privilege.

"But we'll get to that later.
       The point I want to make here
       is that we now face
              a challenge and a possibility.
       The challenge is this:
              that all of us humans have,
                     deep within us,
              strong passions
                     and potential for violent
                     acts that you and I

would rightly consider
immoral, wrong, even evil.
The possibility is this:
that all of us humans have,
deep within us,
strong compassion
and potential for deep connection
with each other
and with the Infinite.

"Ancient Mind is enmeshed within
the Web of Life,
while Modern Mind
struggles against it
and sees itself as separate from
the Web of Life.
Yet both remain fully human,
including the shadow side
of being human."

"So there isn't anything inherently flawed in
me or my ancestors?"

"Of course not, although it
bears repeating—
this does not let Modern Mind
off the hook in facing up to
and taking responsibility
for what it has done.

"This brings us back
to your question about
'the most meaningful lives
that we could ever hope to live.'
I'll start with a story
where Ancient Mind
and Modern Mind meet.

"There was once a village
    deep in a forest, removed
    from the civilizations of Modern Mind,
    yet not so remote that Modern Mind
    was unable to make contact.

"The villagers were struggling
    with malaria and other ailments
    that were new to them,
    because of their recent contact
        with the societies of Modern Mind.
    There were some doctors of Modern Mind
    who felt compassion for the village's plight
        and sought to help its people.
    These doctors told them of the wonders
        of the medicine of Modern Mind,
            and while they took pains
            to not let them show,
            they held disparaging
            attitudes toward the
            healers of the village."

                  The Seeker gave a small snort. "Typical."

The Sage frowned.
    "Perhaps, but don't
    be so quick to judge . . .
    let me finish the story."

                  "Oh. Sorry."

"Now, then . . . the doctors
    were indeed able to treat
    some of the villagers,
    until they encountered
    a young man named Paulo,
    who was extremely lethargic.
    They couldn't isolate any problem
    and became uncertain as to

how to proceed.
What do you suppose they did then?"

"I imagine they kept trying different drugs,
kept guessing at what was wrong."

"They did, but eventually one
suggested sending him
to the village's healers.
Even though it did not fit
within their stories,
the doctors realized that
they had nothing to lose.
They swallowed their pride
and sent Paulo to the healers,
who were in fact able to cure him
using their traditional ways
and knowledge.

"This occurred several times.
Soon, the doctors and healers,
overcoming their mistrust of one another,
began to talk.

"The doctors realized that they
had much to learn from the healers,
who had vast knowledge
regarding what plants and therapies
to use to treat many of the maladies
the villagers encountered.

"The healers, too, realized that
the once-proud doctors did indeed
bring useful knowledge and techniques:
ways to set broken bones,
mosquito nets to prevent
the spread of malaria,
modern drugs for those
who fell sick with certain diseases.

"In time, the village found itself
     with two clinics,
          the medicine of Ancient Mind
               side by side with
          the medicine of Modern Mind.
     The healers and doctors
     sent patients back and forth as needed.
          They learned from and taught one another,
          even bringing some young adults
          from the village into each clinic
          to learn the ways of both Minds.

"What might we learn from this story?"

                    "That one way is not always better than the
                    other, that Modern Mind and Ancient Mind
                    both have something to offer," replied the
                    Seeker.

"And they have the potential
     to be *stronger* working
     together, as long as each
     approaches the other with
     respect, patience, and humility.
     It is this example
          that illustrates the challenge
          and opportunity for
          the evolution of humanity,
          and in fact may be
          the only path that humanity
          can successfully navigate through
          the challenges it now faces."

                    "So you're saying neither Ancient Mind nor
                    Modern Mind can stand alone," said the
                    Seeker.

"Not without enormous risk.
     Humanity has crossed a vital boundary,

a crossing that neither Ancient Mind
nor Modern Mind can ignore
without the gravest of consequences
    to each.

"Even though some of your recent ancestors
    enveloped, destroyed,
and enslaved mine for centuries,
    the two stories,
Modern Mind and Ancient Mind,
    have begun to see one another
across this chasm of violence and pain;
    some have begun
    to open one another's hearts
    to begin to heal these deep wounds—
    and by doing so, they also help
      one another face
        a planetary crisis.

"The way forward is not to go back;
    the way forward is
toward an emergence out of the two Minds,
    a powerful union
of Ancient Mind with Modern Mind,
toward the emergence of a third Mind,
    which I call 'Planetary Mind.'"

          The Seeker leaned forward. "Planetary
          Mind—so that's like a combination of
          Modern Mind and Ancient Mind."

"It's more than that.
    It is a new worldview,
    a new story that does indeed
    encompass the lessons learned
      by both,
      lessons from the seen world
      and the unseen worlds and forces

that flow just beyond
our everyday perception.

"Yet it goes beyond this integration.
    Just as the emergence of Modern Mind
        was unpredictable,
    I'll wager that there are aspects
        of Planetary Mind that we could
            never predict as it emerges.
    In any case, this new, emerging story
        has the potential to help humanity
        navigate the perilous waters
        it has created for itself.
    Certainly bringing Planetary Mind into being
        qualifies as some of
        the most meaningful work
            one can engage in."

                    "Why the word 'Planetary'?"

"Because it acknowledges
    the deep connection
    that all life has with the cosmos,
        but also acknowledges
        the day-to-day realities
        of living on a finite Earth.
    'Planetary' also emphasizes that
        our actions have planetary effects
        and that we must be mindful of what we do
            if we are to survive."

The shadows of the trees stretched across the
meadow. The Sage rose.

"Let's walk back, and I'll finish our time
    with the Story of Planetary Mind."

REUNITING WITH THE INFINITE

1.

It has come to pass, then, that Ancient Mind
    and Modern Mind
    stand together at the abyss—
        at the verge
        of the sixth collapse of the Web of Life,
            the Sixth Great Extinction,
        and thus their mutual destruction.

After generations of injustice—and liberation;
    centuries of exploitation—and innovation;
    millennia of violence—and peace-making;
        they stand together,
            wary—and hopeful,
        searching for a way forward
        beyond the abyss.

2.

Modern Mind can join
    with Ancient Mind;
    Ancient Mind can join
        with Modern Mind,
            and together,
        they can create a new story:

a story that brings together
    the logic and fire
        of Modern Mind
    with the intuition and heart
        of Ancient Mind,
    transcending them both
        with a deeper understanding
        of humanity's place in the Infinite,

creating something never before seen

in the history of humanity,
nor in the life of Earth herself.

3.

Ancient Mind knows
      its place in the Universe
      through the lens of deep intuition,
            telling a story of connection
            and knowing without words;
      Modern Mind understands
            its place in the Universe
      through the lens of deep logic,
            telling a story
                  of technological triumph
            but also
                  of profound disconnection
                        and separation,
                  of knowing only with words.

Both stories stand at the threshold
      of a new leap of creation.

This creation has already begun,
      as Modern Mind has discovered
            that the Universe is
            not logical in the way
      it has come to understand logic,
            that the Story of Modern Mind
            is hollow, limited, incomplete,
            and deeply unsatisfying.

This joining has already begun,
      as Ancient Mind has learned
            the value of some of the discoveries,
                  technologies, and ways
                        of Modern Mind,
            and has recognized, as well,
            that the liberation of Modern Mind
                  is bound up with its own liberation.

Ancient Mind has grasped the need
to face the future together
with Modern Mind.

Both have begun to see the potential
for the two Minds to unite.

And from this realization,
a new, emergent story has begun—
the Story of Planetary Mind.

4.

Planetary Mind
understands that which can be said
and knows that which cannot be said.

Planetary Mind
both knows in an intuitive sense
and understands in a logical sense.

Planetary Mind understands and knows
humanity's connection with
and place within the Infinite,
recognizing that the Mysteries of the Infinite
cannot always be explained,
nor must they always be—

that there are truths that lie beyond
the literal and logical.

5.

Ancient Mind stands in respect and awe of the Web of Life,
thriving upon its blessings and bounty.

Modern Mind stands in fear and awe of the Web of Life,
using technology to try to dominate it.

Planetary Mind stands
in awe and respect of the Web of Life,
thriving upon its blessings and bounty,
mindfully using technology to dance in concert

with the lessons of the Web of Life
and the mysteries of the Infinite.

6.

Planetary Mind creates stories about the Infinite
    and the origins of the Universe—
        seeing itself within the Web of Life,
        adapting its stories for many cultures.
    Planetary Mind knows that it
        writes the stories of creation itself
            and that stories shall change
            as greater understanding
            and knowledge
                come to light.

And the overarching Story
    that Planetary Mind is now creating for itself
        is this:

The creative force of the Infinite
    flows throughout and embraces the Universe,
        whole and complete:
            ever changing,
            ever creative,
            and ever destructive,
    as the Infinite plays and dances—
    as It seizes catastrophe to create anew—
    as It yearns for ever greater order,
        ever greater complexity,
        ever greater consciousness,
        and ever greater elegance—
    as It drives Itself toward life.

We are thus made in the likeness of the Infinite.

We are this creative force of the Infinite,
    manifest in the form of a human being;
    we are an integral part of
        the ways

of the Universe
and are subject to the lessons
of the Web of Life.

And along with the tiniest particles
flashing in and out of existence,
along with the most enormous
clusters of galaxies stretching
millions of light years across,
we, too, belong to the Infinite;
we, too, are a sacred expression
of the Infinite.

It has always been thus; we never were separate.

We can awaken to a deeper communion with Earth,
know and love our place within her Web of Life,
rejoice in her,
and share in her bounty.

And we do so with gratitude, respect,
and humility,
for we know that while we have learned much,
there is still much to learn
as Planetary Mind awakens within
the Holy Universe.

7.

Planetary Mind understands and knows,
and with intention, shapes
its stories,
its cultures,
its societies
to be in harmony with the Web of Life.

Planetary Mind understands
it is inseparable from the Infinite,
and, as is the Infinite,
it is both the painter
and the canvas being painted upon.

Planetary Mind knows
   there is not a creator
      separate from the created:
         there is only creation—
      flowing throughout the Universe,
        throughout Earth,
        throughout the Web of Life,
    and throughout each one of us
      as we stand in awe
        of the starlit
          sky.

We are the Infinite

   and the Infinite is us.

The Seeker and Sage arrived back at the
coffee shop. They stopped and stood silently
in front of it.

The Sun shone orange through the leaves
of the trees and the breaks in the buildings
across the street as afternoon turned
toward evening. Cars passed carrying their
passengers home toward dinner.

"You're probably feeling
   a bit full right now,"
   remarked the Sage.

        The Seeker chuckled. "Yeah . . . yeah, you
        could sure say *that.*" He reflected for a
        moment. "So you're saying that Planetary
        Mind already exists, then?"

"It is coming into being
   as we tell the story,"
   said the Sage.

"There are those
       raised within the story
              of Ancient Mind,
       who see the value of what Modern Mind
              has to offer
and who recognize the need
       to join with Modern Mind;
some of these people
              even assert that a coming together
              of the two Minds has already begun,
                     in fulfillment
              of their ancient prophecies.

"There are also those
       raised within the story
              of Modern Mind
       who have a deep,
       inexplicably strong impulse
       to transcend
       the story of Modern Mind,
              who are unwilling,
                     almost unable,
              to continue with the story
              of Modern Mind."

The Sage looked at the Seeker.

"I'll wager that you're part
       of this second group,"
              she said.

                            "Huh," replied the Seeker. He looked away,
                            thinking. "Maybe so."

The Sage smiled.
       "This is a lot to take in.
       Just sit with this story for now,
              and we'll see what comes up
              in our next conversation.

Until then,"
    she said, giving him a hug
before walking away.

        The Seeker watched her turn the corner, and
then sat down at one of the tables outside
the coffee shop. In spite of being hungry,
he sat for quite a long time, thinking. He
noticed that the sense of unease so familiar
to him had lightened a little.

PART II

THE GREAT TRANSFORMATION

Two paths through the Great Transformation
   now lie before us—one grim, one graceful:

Will we travel along the first path,
   desperately clinging to our old story,
   as a tyrant clings to the crumbling walls
      of his corrupt and dying empire?
Will we deny the reality before us—
      ensuring that our children
      will inherit a Web of Life
that has collapsed, leaving them, at best,
      with a diminished and disabled world,
   if not indeed stranded and starving?

Or, raising our heads above our denial,
   will we travel the second path
   and recognize the enormity
      of the dangers before us?
Recognizing these dangers,
         will we see that we must act?
      Seeing that we must act,
            will we then indeed act
            and squarely face the challenges
      of this Great Transformation,
      to bring forth the new story
            of Planetary Mind?

— THE SAGE

THE JOURNEY THROUGH THE GREAT TRANSFORMATION

"So we'd like to offer you the job," the supervisor said with a smile.

The Seeker was startled. "Wow. Uh, thanks . . . I didn't expect this just now." He was glad to get to this second interview, because jobs in construction were scarce. But while he was sure he could do the work—he'd done project management in his last job—he still worried that his heart wouldn't be in it.

"The owner, she really liked you; you've got a good attitude and that's important to us, and you're more than qualified."

The Seeker smiled in spite of himself. "Well, thanks," he said again. "Can we talk about salary?"

They negotiated the salary, benefits, time off, and the starting date.

"Any chance we can put this down in writing, make it official?" said the Seeker, after they were finished.

"Oh, sure, sure—that's fine. I'll send it along
by tonight. Sound good?"

> "Great," said the Seeker, rising and shaking
> hands. "I'm looking forward to telling my
> wife the good news."

> As he walked out of the shop, the knot in his
> stomach eased but didn't quite disappear.

"Congratulations!"
    exclaimed the Sage,
    clapping her hands together,
        smiling.
    "You certainly found one
    quickly enough, yes?"

They were sitting in blue fold-up chairs
that the Sage had set up on a small clearing
on a hillside overlooking the city, which
thrummed with the activity of unseen
thousands.

> "Yeah, a lot easier than I expected, especially
> these days. My wife thinks I should take
> the job, at least for now, and I see her point.
> We have bills to pay and everything. This'll
> relieve the stress. Some of it, anyway."

The Sage narrowed her eyes.
    "You've found this job,
        relieved some stress;
        yet I gather
        that there's still much for us
    to explore together, yes?"

> "Oh, yeah, I definitely want to talk more. I
> mean, you were right; this job isn't going to

solve everything going on with me. I need
to talk more about the story. You know,
Ancient Mind, Planetary Mind . . . what
does it all mean, what does this mean to
me?"

"Ah, well—I would certainly love
to discuss that with you,
as long as you're clear that
you're not expecting answers
from me as if I were
an 'all-knowing sage.'"

"Well . . . didn't you give me a bunch of
answers last time we talked?"

"I gave you a *story*—
a way of looking at the world
you might find helpful.
Do you recall what I said about
that story, right before
I told it to you?"

"You said something about how it's a young
story, that it changes."

"How does it change?"

"It changes as we learn, as we get more
information."

"As we learn and get more
information about . . . ?"

"Umm—well, the Universe and the Web of
Life."

"And?"

The Seeker thought. "And about ourselves,"
he said after a moment.

"Yes—and *you* are part of
    that discovery. Certainly
    you'll meet teachers
    of all kinds, but you, too,
    have your own discoveries,
        large and small,
    to add to the story."

                "But I'm no expert. Aren't you the expert?"

The Sage shook her head.
    "Maybe I've been thinking
        about these things longer
            than you have;
    maybe I've learned a few things
        that you might find useful,
            and I'm happy
        to share my worldview with you.
But never, never forget:
    I, too, am as much of
    a seeker as you are,
        as anyone is.
Even if I happened to have
    vast amounts of experience,
    you would not serve yourself
        by holding me,
        or anyone, for that matter,
        as some kind of special authority.

"In some ways, I'm sure,
    I'm more of a neophyte than you.
    Any word from others you might rely upon,
    written or spoken,
        is at best a guide,
    and at worst
        a dangerous surrogate."

"Why have these conversations, then?"

"I love the practice."

The Seeker laughed.

The Sage laughed with him. Presently they
fell silent, looking out at the city below.

"And perhaps you, too, will continue
    to find these conversations
    useful and enjoyable,"
    she said after a while.
    "That's for you to decide.

"But tell me this:
    who has the final authority
    over how you are in the world?"

"I guess it's me," said the Seeker.

"Spoken with such confidence,"
    chuckled the Sage.

The Seeker frowned.

"Look, you're not alone
    in that struggle.
    We've been trained
        to see ourselves as separate,
    trained to see authority
        as entirely *outside* of ourselves
        whether we respect
    or rebel against such authority.

"Those of us raised within
    Modern Mind's societies
    have been taught since the cradle
        that answers lie on the outside,
        and it's hard to let go
    of that story, that craving."

The Seeker nodded and sighed. "Yeah,
you're right. It would be nice to have the
answers handed to me."

"Beware of this desire.
    Beware of those who are unable,
    unwilling—even afraid—
        to bask in the mystery.
    Beware of anyone or any book
    that has explanations for everything,
    that claims to have all the answers
        handed down from on high.

"Perhaps such claims might have
    been useful to humanity in the past.
    But I believe they don't serve—not you,
        nor the whole of humanity.
    It is high time we acknowledge that all of us,
        as children of the Universe,
            as the Universe itself,
        are *ourselves* creating our stories.

"Now, as we learn to see each other
        as *connected* selves,
    and learn how to reap
        our collective intelligence,
    this dichotomy
        between authorities and oneself
            starts to dissolve.
    We learn of a way
    to tap into our *shared* authority,
    that arises out of skillful
    dialogue and interaction.

"But even in these inquiries,
    you are still the authority
        in your own life.
    You would do well to also learn
        to connect with that wordless

knowing within you
and pay attention to the callings
     of your spirit,
for it is upon *this* that rests
     your final authority."

                    "Hmm." The Seeker mused. "I know what
                    you're saying, but . . . well, I guess I'm just
                    not used to seeing myself that way."

"Let's go back to
     what you said earlier,
          about wanting to talk more.
     This story has struck you;
          how has it done so?"

                    "So many things . . . well, for instance, this
                    idea about Modern Mind: that we're not
                    flawed, we're not a mistake—is it possible
                    that the Universe, the Infinite . . . that
                    everything's all really just fine, and that it
                    will all work out in the end?"

The Sage chuckled.
     "'It all works out in the end';
          wouldn't that be nice!
     But the thing is,
          there's *never* an end.
     The story of the Universe
          goes on and on;
          the Web of Life
          is never finished.
     However, a story of a Universe
          that knows what it is doing
          might be a useful
          image to play with."

                    The Seeker shook his head. "But that
                    explanation bothers me; somehow you

could turn that into the idea that we don't
have to do anything, and that feels just
completely *wrong*."

The Sage narrowed her eyes
    at the Seeker. "I'm curious,"
        she said.
    "Have you always felt
        this passion?"

                    "What passion?" said the Seeker.

The Sage threw her head back
    and laughed heartily.
    "What *passion?*
    The same passion
        that brings you to this woman
        and insists on asking
        these enormous questions,
    that is not content
        to go along with life as
        Modern Mind presents it to you!
    Yes, your *passion!*"

                    The Seeker felt his cheeks redden. "Well, I
                    guess . . . I don't think of it that way."

"I guess you don't,"
    said the Sage, playfully.
    "It might be good to
    recognize it, yes?"

                    The Seeker thought for a moment. "There
                    was this one time, when I was a kid." He
                    stopped, remembering.

The Sage cocked her head.
    "Tell me about this.
    Close your eyes
        and remember."

He closed his eyes, drew in a breath, and let
it out.

◆

It's the afternoon. I'm about eight years old
  and I'm wearing my jeans and a shirt I like
  with blue, yellow, and red horizontal stripes.
  I'm in the living room with my brother and sisters,
      sitting on the old gray couch
      watching television.
  It's some kind of nature show,
      and I'm horrified;
  it's showing the desecration of the land,
  telling of the extinction of species.

One scene: of ranchers shooting prairie dogs,
  blowing them to pieces with rifles,
      because the animals
      are considered pests.
  I'm shocked at how cavalier the ranchers are;
      something's horribly wrong
  with their thoughtlessly killing
      these animals.

Another scene: a flock of large, elegant cranes
  flying across an Asian landscape,
  and the announcer intoning,
  "These are the last eight of their kind.
      When they die, their species
      will be lost forever."

And another scene, and another . . .

I lose it. I start crying. Not just crying, but sobbing.
  I can't believe this destruction is happening,
  that it's being allowed to happen.
      It feels so deeply wrong.

Somehow, I end up in Dad's arms,
>   even though I know he hates it when boys cry,
>   but I'm practically hysterical.

Poor Dad—he must have been
>   scared himself, I was crying so hard,
>   as if I had accidentally
>   shredded my fingers with his power saw,
>   or snapped my leg, falling out of one of the trees
>   next to the creek near our house.

Something about
>   the mindless killing of animals,
>   about the cranes disappearing forever—
>   it was as frightening as losing my parents forever.

My sister came over, started to tease me,
>   but Dad shushed her away.

When he finally got out of me what had happened,
>   he was bewildered.
>   I remember him saying,
>   "My God, son,
>   of course, this is something
>   to be concerned about,
>   but certainly not something
>   worth being so upset over."

But I didn't believe him, couldn't believe him.
>   I really didn't have the words,
>   but if I did, they would have been
>   something like,
>   "Oh, yes, yes it is . . .
>   It certainly is worth it . . .
>   It most certainly is something
>   worth being so upset over."

◆

The Sage sat looking across at the city below.
    "Mmm. Mysterious, isn't it,"
she said after a moment.
        "Where this passion springs from;
        why you, and not your siblings—
        where does this caring come from?"

She shook her head, smiling.
    "Such a delightful mystery."

She turned to the Seeker.
    "Fortunately, we don't need
    to fully understand the mystery,
    only to see that it's there.
        And it is always good
        to find a kindred spirit."

                    The Seeker raised his eyebrows. "Kindred
                    spirit . . . well, okay. Some of my friends
                    think my wife and I are a little crazy."

The Sage laughed.
    "Do you think they're right?"

                    "I wonder." He paused. "Well, no, but it sure
                    feels lonely, sometimes."

She let out a breath.
    "Well, then . . . this is a good place
    to continue our conversation,
    with such passion in mind."

CATASTROPHE AND CREATIVITY

"It is this passion that will lead you
    and lead all of us, as we face
    the Great Transformation,
    which is happening now
    over the entire globe."

        "Great Transformation," said the Seeker. "I've
        never heard about that."

"I'm not surprised—
    it certainly isn't mentioned in our media.
    But just as early life radically changed
        the Web of Life
        as it came into being
        three billion years ago,
    humanity itself
    is now doing the same.

"The Great Transformation
    is the sum of these enormous,
    epic, global changes,
    and it is already underway.

"Unfortunately, these changes
    are profoundly diminishing
    the ability of the Web of Life
    to provide for us
        and all of its children."

        "You're talking about climate change, right?"

"Oh, much bigger than that.
    You can look at these crises
        through three lenses:
    social,
        spiritual,

and ecological—
and they're all related.

"Socially, we have enormous
accumulations of wealth,
which, instead of bringing prosperity to all,
have brought riches to the few
while impoverishing billions of others.
And those of us who are well off
have the power to change,
but we don't, in part because
we're isolated from
the results of such inequity,
and in part because our story
allows it to continue.
I doubt humanity can endure
under the strain
of such deep unfairness.

"Yet we—the minority
who lead lives of comfort unimagined
only a few decades ago—
we don't know what fulfills us.
Modern Mind's consumer worldview
tells us to buy more and more,
but this is deeply unsatisfying,
and we suffer from mental distress
and spiritual starvation
as deep as any physical poverty
found on the planet.
We can hardly go forth
into the future
with such bleakness
and emptiness
within so many souls.

"And these social and spiritual crises
are directly related to

the same worldview
that damages the Web of Life.

"We've cut down forests for farms,
    over-fished the seas,
    mined minerals and fossil fuels,
    churned out piles of consumer junk,
    and built cities sprawling
    across the landscape, causing
    not just catastrophic climate change,
    but even more ominously,
    we've triggered
    the Sixth Mass Extinction,
        which I'll wager
        is close to your heart,
        given the story
        you just told me."

                    "Wait . . . you said something about that
                    before in the story you told me, but what
                    exactly do you mean by the 'Sixth Mass
                    Extinction'? We haven't been hit by an
                    asteroid, have we?" asked the Seeker.

"No, but species *are* disappearing
    from the Web of Life
    in unprecedented numbers,
    due to the polluting and
        destruction of their habitats,
    exploitation of species,
    and the introduction, by us,
        of species and diseases into habitats
    in which they did not evolve,
        something we did both accidentally
        and intentionally.

"As I pointed out in the story of
    the Emergence of the Web of Life,
    a catastrophe of this magnitude

has happened only five times before on Earth,
　　the last one being sixty-five million years ago,
　　when the dinosaurs became extinct.

"And now, the rate of extinction
　　has increased such that,
　　if unchecked, it is likely
　　that half of all species of life
　　will be extinct within
　　　a generation."

　　　　　　　　　　　"Half?"

"Yes, half."

　　　　　　　　　　"That's . . . unbelievable."

The Sage sighed.
"Yes, unbelievable.
　　　I find it hard to accept
　　that the sacred, talismanic,
　　　　and heartbreakingly loved
　　　companions of ours
　　are about to disappear forever.
　　　They will not return."

　　　　　　　　　"My God." The Seeker stared out at the
　　　　　　　　　city below, stunned with this news. He sat
　　　　　　　　　silently for a while. He watched a flock of
　　　　　　　　　crows flying overhead, calling raucously,
　　　　　　　　　almost playfully, to one another.

　　　　　　　　　After a moment, he asked, "We're facing
　　　　　　　　　something that hasn't happened in sixty-five
　　　　　　　　　million years?"

"Yes, we are."

　　　　　　　　　The Seeker was silent again. "I knew that
　　　　　　　　　things were bad," he said finally, quietly,
　　　　　　　　　"but I had no idea they were *that* bad."

The Sage nodded.
    "You are not alone, I'm afraid.
    One would think that these crises
        would be of paramount concern to all,
        yet it is frightening how oblivious
        Modern Mind is to these facts—
    especially when there are scientists
    who assert that humanity itself
    will not survive this extinction."

                "But why in the world . . . why don't we
                know about this?" said the Seeker. He got
                up from his chair and walked a few yards
                toward the edge of the clearing. He began to
                pace.

"Once again, it isn't readily apparent to us
    as we go about our lives;
        the feedback loops aren't in place.
    We don't *see* the pollution,
        the habitats disrupted and destroyed,
        and the species disappearing.
        Individually, people of Modern Mind
            are unable to perceive such
            enormous changes.
    We can only perceive these crises
        through collective scientific work,
        but we haven't yet fully learned
        how to *respond* to them.
    And because it's not in our faces
        every day, it's easy for us
            to live in denial.
    We also don't know how
        to comprehend such colossal tragedies,
        how to *feel* such immense losses.

"Perhaps, too, because these facts
    threaten the story of Modern Mind,

any suggestion that its story
may be a root cause of our plight
    invites disbelief,
    rejection, or even anger.
Certainly our media have ignored,
    downplayed, and even
    discredited these facts."

The Sage nodded toward the city below.
    "Sometimes, when I contemplate
the enormity of it,
I myself don't quite know
    what to think."

        "We're living on a planet that soon won't
        support us anymore," said the Seeker. His
        voice strained with agitation.

"Hold on," she said,
    raising her hand.
    "There's an important point
        that I return to
        when I find myself becoming
        wrapped up in fear.
    Think back again to the story of
        the Emergence of the Web of Life.
    What happened when
    some of the earliest beings
    found a way to use sunlight
    and water to create food
        for themselves?"

        "Um . . . I don't remember."

"'They gave off oxygen,
    a poisonous fire,'"
        quoted the Sage.

        "Oh, right . . . when some of the beings gave
        off oxygen, it was a huge problem."

"Dysfunctional, you might say.
     It was in fact catastrophic
     and threw the Web of Life
          into turmoil.
     Remember what happened then?"

> The Seeker shifted his weight, thinking. "I
> remember . . . another creature figured out
> how to use the oxygen, and then it wasn't a
> problem anymore."

"There's a little more to the story.
     The being that unlocked
          the power of oxygen
     itself became more powerful,
     with the potential for destroying
     and devouring any other being.
     But what happened then?"

> The Seeker frowned. "I remember it going
> hunting," he ventured.

"'Competition begat cooperation,'"
     said the Sage,
          quoting again.

> The Seeker nodded his head, remembering.
> "That's right—one of the beings it attacked
> didn't die, and they figured out how to work
> together."

"They became a new being
     unto itself, surpassing
     the potential that each had
          on its own.
     You could say that the one being
          that harnessed oxygen
          was indeed more competitive.
     But it was the cooperative joining
          of the two that biologically formed

a much more complex cell,
which in turn created the potential
for multi-cellular creatures.
This cooperative alliance allowed
for greater complexity of life.
Without it, we literally
would not be here."

"You have to have both for things to evolve."

"Yes—and this is a vital lesson:
cooperation is as much
a facet of evolution
as competition, perhaps more so.
Modern Mind has ignored
one of the most important lessons
we can draw from the story:
life itself created
dysfunction and catastrophe.
But these crises led to new creativity
through both competition
and cooperation.
Evolution—creativity—happens all the time,
but every now and then it is
spurred on by catastrophe.

"Remember, there's nothing
inherently wrong with us;
we've created and filled a niche
and have been incredibly
successful doing so.
Just like those early beings
who brought forth
the oxygen catastrophe,
we've brought forth
another global catastrophe—
but these crises could very well spur

the evolution and creativity
of Earth and the Web of Life."

> "So you're saying that this is what's
> happening here, that catastrophe will force
> us to change?"

"One way or another, yes;
    like it or not,
        the Great Transformation
        is upon us like a tsunami,
    and we don't have much time."

> "But like you said," the Seeker motioned out
> toward the buildings and freeways below,
> "it's too big, too entrenched, the way we do
> things now."

"Yes, and Modern Mind
    has overstressed the idea of competition;
    this idea certainly has the potential
    to continue to misguide us."

> "Right. And do people really *want* to change?
> It all sounds incredibly impractical."

"In time, the impractical
    will become inevitable,
    and it won't matter
        whether people want to change.
    Humanity is living beyond
    the means of Earth's bounty,
    using one and a half times
    what Earth is capable of
        providing each year.
    It's as if the streams feeding a lake
        can replenish it
        with only two-thirds of a bucket
            for every bucket
        each person now takes from the lake.

The level of the lake is lowering,
  whether or not we notice,
    accept,
        or deny this reality.
If we don't make changes,
  the lake will dry out—
and our civilization will disappear,
just as other civilizations have."

                    "We're headed toward a wall."

"We are. Two paths
  through the Great Transformation
    now lie before us—
      one grim, the other graceful—
as the realities of what we have wrought
become harder and harder to ignore:

"Will we travel along the first path,
  desperately clinging to our old story,
    as a tyrant clings to the crumbling walls
      of his corrupt and dying empire?
Will we deny the reality before us—
    ensuring that our children
    will inherit a Web of Life
that has collapsed,
    leaving them, at best,
      with a diminished and disabled world,
if not indeed stranded and starving?

"Or, raising our heads above our denial,
  will we travel the second path
  and recognize the enormity
    of the dangers before us?
Recognizing these dangers,
      will we see that we must act?
  Seeing that we must act,
      will we then indeed act
        and squarely face the challenges

of this Great Transformation,
to bring forth the new story
of Planetary Mind?"

The Seeker sat heavily back into his chair,
and looked silently out from the hillside.

"Tell me what you're thinking,"
said the Sage after a while.

"It's just a lot to take in," said the Seeker. "I
mean, if what you're saying is right, we're in
big trouble."

The Sage chuckled.
"Indeed we are!"

The Seeker frowned. "Why are you
laughing?"

"To drive home a paradox:
we are not in charge,
we can't predict the future,
and the Infinite knows what It's doing . . .
at least It *acts*
as if It knows what It's doing.
We've seen this all along."

"But I don't get it! We're talking about huge
crises, right?"

"Yes, we are,
and yes, it's important to take
these crises seriously,
and especially important
to learn the lessons
they bring to us and take action.
It's like having pain in your body;
within the pain, within the crisis,
lie sources of information
about what to pay attention to.

"But taking them *so* seriously?
How *useful* is such an attitude?
And where is the fun in that?
How much does it serve
    to cling to *any* story about the future,
    especially a fearful and anxious story of
'we have to save the world'?

"As I contemplate the death of species,
    as I contemplate the billions
        in abject poverty,
    as I contemplate the sad,
        empty hearts of the human beings
        we insult by reducing them
        to mere 'consumers,'
I find it necessary to remind myself
    that if I want to change things,
I had best start with changing
    my *own* heart,
to allow it to break, yes,
    but not to break down—
    instead to break *open*."

The Seeker said nothing. Tears filled his eyes,
and he looked away.

The Sage sat quietly, waiting.

"I don't know. I'm . . . I guess I want to feel
hopeful or upbeat, but . . ." He stopped.

"You're going back and forth in a way that's
really confusing," he blurted.

"In one moment, you're saying we're at the
brink of destruction, and in the next, the fix
is already in motion. In one sentence, you're
talking about millions of years, and in the
next, that a single generation can either blow
it all or make it all okay!"

"Upsetting, yes?"

> *"Yes, goddammit!* Especially when I think of the mindlessness and ugliness of it all—and how am I supposed to raise a child, when she might not even have a planet to live on? I feel helpless!"

"Take some time.
    Feel this upset in your body."

> The Seeker sat, looking down at the ground. He closed his eyes as the tears gathered and fell. He put a hand to his face, resting his elbow on his knee.
>
> After a bit, he wiped his face on his sleeve, and took a deep breath. He looked at the Sage and then stared out at the city.
>
> "It just gets me right here," he said, pressing a hand just below his sternum. He fell silent again, rubbing his gut.

"It's a delicate balance,"
    said the Sage, quietly.

"You want to feel the upset,
    but not be swept away by it.
    Collapsing into a depression
        won't serve, and neither will
    numbing yourself nor making yourself
        sick with worry."

> The Seeker dropped his hand into his lap. "Sometimes I wish I *could* numb myself. I wish I could just go back to how it all was and not think about any of this." He sighed and sank back into his chair.

"These are enormous times to live in,"
        continued the Sage.
    "While it doesn't serve to latch on
        to stories about the future,
            because no one can be certain
            about what's going to happen,
        we certainly can't ignore
            what science is telling us
        and what our hearts are telling us.
    It's challenging for anyone
        who opens his or her heart
    to what is becoming increasingly obvious:
        what we do within the next few years—
        and we indeed may only have a few years—
            will have a profound
            impact upon our descendants
            for hundreds of generations.
    That's how enormous the crises
        of the Great Transformation are.
            We are not at the eleventh hour,
                we are at *the* hour."

                    "Hah. No pressure," said the Seeker.

The Sage smiled.
    "No, none at all!"

She took in a deep breath and let it out.
    "Let's take these one at a time.

"First: the story of Modern Mind
        cannot continue; we're going
        to have to change, yes?"

                    "One way or another."

"Second: catastrophe inspires creativity.
        We've seen this
            with each Great Extinction.
        Over time, evolution *demands*

greater and greater complexity,
        higher levels of order and cooperation,
        and ingenious solutions to problems—
                and has done so
        ever since life emerged.

"You can even see this
        with human catastrophes,
    whether it's people
        making their way after
        a flood or an earthquake,
    or a person confronting a serious illness.
    There's no guarantee
        that things will work out,
    but such catastrophes can
    bring out the best in people
        and serve as wake-up calls, yes?"

                        "Yeah, but they can bring out the worst in
                        people, too. And a lot of the time people
                        don't wake up until it's too late."

"I won't argue those points, and I pray
        that the catastrophes we face
        will not have to go far beyond
            some critical point of no return
            before we do awaken."

                        "If we haven't *already* passed the point of no
                        return," said the Seeker.

"In some ways, we already have.
    We're faced with
    a planet that is already changing
        and will continue to change,
    even if we suddenly reverse our course.
    But, when given a chance,
        the Web of Life can rebound
        with astonishing quickness.

There's a good possibility that
　　we can help the Web of Life
　　　　heal what it can,
　　while we flow with the effects
　　　　of these crises."

　　　　　　　　"Well, okay. I think I get that."

"Third, and most hopeful:
　　remember that we *belong* here,
　　　　we are as much a part
　　　　of the Web of Life
　　　　as any being in the Universe.
　　Yes, we're no more precious
　　　　than any other species,
　　and can easily vanish from
　　　　the Web of Life,
　　　　　　as all species eventually must.
　　But when we choose the second path,
　　　　choose to bring forth Planetary Mind,
　　we align ourselves with
　　　　a stupendous, deep, and powerful drive
　　　　　　inherent within the Infinite.
　　We align ourselves with
　　　　a cosmically ingenious and mysterious drive
　　　　　　toward complexity, order, and beauty,
　　　　　　this drive toward life
　　　　seething through the Universe
　　since the Beginning of All Beginnings.

"It's as if we're attending
　　a birth; the baby's coming,
　　　　ready or not.
　　It will take work; there will be pain,
　　　　even blood,
　　　　and no guarantees
　　　　that things will work out,
　　but deep down, the bodies

of mother and child
know what to do,
and the Infinite
will *not* be denied."

The Seeker pondered. "I didn't think of it
that way." He picked up a stick from the
ground and twisted it between his fingers as
he rolled the thought around in his head.

"No guarantees, though." he said.

The Sage nodded and shrugged.
"I've also heard it argued that
while we *are* destroying
the Web of Life's
ability to sustain us,
perhaps driving ourselves to extinction,
we're *not* destroying the planet, not entirely.
The Infinite has Its ways;
perhaps Planetary Mind is a seed that
is ultimately not destined to flower.
Earth will then continue
on her own creative dance
through the galaxy
and discover her destiny
without us."

"That sounds sad," said the Seeker.

"It's sad to me, too.
But I've even heard the idea that
humanity is *supposed*
to cause this extinction,
so that the Web of Life
can regenerate anew,
and Earth can move on
to her next venture."

"Do you *believe* that?"

The Sage shook her head.
   "Not for a moment,
      for there are also no guarantees
      that the Web of Life *will*
      be able to regenerate.
   And what if the Universe somehow *needs* us
      to help create a new level of evolution,
      to *consciously* create new levels
      of cooperation and complexity?
   In any case, I've chosen to not sit idly by
      and watch as Modern Mind
      tears the Web of Life to pieces,
  to instead work to change
      the story of Modern Mind,
         to learn and teach what I can
      about how to live in harmony
         with the Web of Life.

"And even if our civilizations
         endure a catastrophic collapse,
  we're a resilient species;
  perhaps some survivors will continue
      humanity's journey—
  and I certainly believe we can pass along
      the story of Planetary Mind
         in some way to them.

"But that aside, my hunch is that the odds
      are in favor of humanity's evolution.
   I doubt anyone can imagine
      what possibilities await us
      as Planetary Mind awakens,
  just as the monarch caterpillar,
      who thinks the Universe is a single tree,
      cannot imagine the vast distances
      she will travel when she emerges
      from her cocoon.
         That sounds fantastic.

To me, the choice is obvious:
    even without guarantees,
       even with the possibility
     we may already be too late,
which path calls me to my higher self?
  Which path sounds more *fun?*"
    she asked, smiling.

          The Seeker shook his head. "I know you say
          to keep it light and all, but I still don't see
          how you can act so *happy.*"

"I might just as well ask:
   why are *you* so *unhappy?*
   Or, better, why do you
     make yourself so unhappy?
Certainly I've given you a lot to take in,
   but even so, I've sensed
   this proclivity in you."

          "I don't—" began the Seeker. He stopped
          and looked away.

"Yes?" said the Sage,
   raising her eyebrows.

          The Seeker sighed. "Okay, okay, maybe
          you're right, but I'm not quite sure how to
          get out of it." He let out a breath. "Look,
          can we break here? My mind's feeling . . .
          full, I guess."

"And your heart, too, perhaps?"

          The Seeker dug at the ground with his foot.
          "Yeah, heart, too, I suppose. I just need to sit
          with it all for a while."

"Of course. For now,
   don't fight your
     thoughts and feelings,

but watch them,
as you've been doing.
Keep watching your stories.
And let more than your mind
    sit with this.
Let your heart and gut
sit with it, too."

                     "Okay," said the Seeker. "I'll give it a try."

"That's all I can ask,"
    said the Sage, smiling,
    as she rose and folded up her chair.

THE WORK OF THE
GREAT TRANSFORMATION

"Look, I'm sorry I didn't do the dishes, but
I lost track of time, and I had to get Ursula
to practice. It's not a big deal. I'll do them
now!" exclaimed the Seeker's wife.

"It's not that," said the Seeker. "It's just, you
said you'd do them, then I walk in and the
sink's full of dishes . . . it just throws me, I
hate it when they're not done."

"I know, I know, I'm sorry, but sometimes
I'm going to mess up and it's like that isn't
okay with you."

"But . . . it's just . . ." the Seeker sighed. "I
don't know . . ."

"I don't know, either," she said, exasperated.
"I mean, I get that you don't like it but why
does such a little thing have to be such a big
deal? It's like you *make* yourself so unhappy!"

"I don't . . ." The Seeker stopped, startled.
He looked at his wife, and then looked away.
They stood silently for a minute.

"I guess I do that a lot, don't I," he said, after
a while.

"Yes, you do," she said quietly, picking up a
dishrag.

◆

The Sage and the Seeker were sitting in the
park on a bench next to a playground. Two
children were building sand castles, one light

skinned with long brown hair, and a second,
darker skinned, with rows of colorful beads
woven tightly into her hair.

A man sat at a picnic table close by, reading
a book, glancing up at the children every
now and then.

"She said nearly the same
        thing I did, yes?"
                The Sage laughed
        and clapped her hands.

                        "I don't see what's so great about it." The
                        Seeker frowned.

"Because you *noticed* it!
        You didn't just continue on in
                the same old story.
        That is cause for celebration, yes?
        Go ahead, say it: 'good job!'"

                        "Good job," said the Seeker, unconvincingly.

The Sage cackled with delight. One of the
children turned to look at her as she laughed,
then returned to her sand castles.

"You can do better than that!
        Really think about it;
        realize that you *chose*
                a small sidestep from the
                way you normally dance
                unconsciously."

                        "Okay." He let out a breath. "Good job. Yeah,
                        I guess it was a little better."

"Very uncomfortable
        to say this, isn't it?"

The Seeker chuckled. "Yes, it is." He sat,
thinking. "My wife's also been reading some
things, kind of like what you and I've talked
about. We talked about how our friends
seem to be pretty oblivious, and how we
sometimes feel so alone."

"Ah, I certainly understand *that* feeling.
    But you might find some surprising
        allies among your friends;
    the awakening within yourself
    is also stirring in the souls
        of millions of your
    as-yet-unknown fellow seekers."

"Millions?" said the Seeker. "I don't know . . .
I mean, I've always felt like I'm the odd man
out. If people really knew what we were
thinking, they'd think we're crazy."

"A common feeling among those of us
        who have begun this awakening.
    But you are *not* alone;
        you can seek out those
        who are like-minded,
        so you might offer
        mutual support and inspiration
        as each discovers his or her life's dance.
    We're going to need each other
        as we square up to the challenges of
            the Great Transformation;
        we'll need not one hero,
        all shining and strong,
        but hundreds, thousands,
            even millions.
    You, of course,
        are one of these heroes."

The Seeker looked up, startled. "Me?"

The Sage smiled, shaking her head.
　　"Don't be so surprised!
　　　　Surely you might have guessed
　　　　that our conversations
　　　　might mean more
　　　than the pleasant exchange
　　　　of two little egos?"

　　　　　　　　　"Well, *I* certainly didn't choose to become a
　　　　　　　　　hero," exclaimed the Seeker.

"Oh?" said the Sage.
　　"I notice that *you*
　　　　seem to be pursuing answers
　　　　to these questions."

　　　　　　　　　"Yeah, but I don't recall signing up for any
　　　　　　　　　hero training, either!"

The Sage threw her head back
　　　　and laughed.
　　"I don't mean what Modern Mind means
　　　　by the word 'hero,'
　　　as a solitary individual engaging
　　　　in larger-than-life acts.
　　The heroes I'm talking about are
　　　　ordinary folks working together
　　　　to change the way things are set up
　　　　　in our society—
some working for large institutional change,
　　　　others who are quieter,
　　　　　　expressing their heroism
　　　　　　　closer to home
　　　　　　through their daily acts.
　　We need *both* kinds of these heroes.

"Let me also be clear:
　　no deity or elite group
　　　　chooses such heroes.

You instead choose *yourself*.
   The choosing is done
   by the spirits and hearts
     of those unable
     and unwilling to slumber
   through the catastrophes
     that are now upon us.

"These fellow seekers have already
   created millions of organizations
   working to transcend
   the story of Modern Mind.
   They are the myriad tiny seeds sprouting
     in the cracks within
     the story of Modern Mind.

"It is a movement of movements,
   with no center, no one in charge,
     a diversity of people
   coming together across the globe.

"And know that to transform
   the systems of our societies
   not every person need awaken
     or agree with the change.
   Not much more than two-fifths
     of the white population in the
   colonies of North America, for example,
     actively supported the Revolutionary War
       against England."

                "Yeah, I've heard that it only takes a minority
                of people to change things," said the Seeker,
                "but Modern Mind is so *entrenched*."

"Let's borrow a lesson
     from those who study systems.
   As a system becomes
     more and more unstable,

it reaches a point where
it either breaks down
        or breaks through to
        new levels of order.
This is called the *bifurcation point*.
And here's the thing: at this point
even the smallest fluctuation
can influence the direction
        of the system.
A big red rubber ball
        perched on top of a hill
        might roll down
        in any direction,
        with the slightest puff of wind.
It's at this point when things are
        especially unpredictable."

                "The more instability, the easier it is to
                change the system, then?"

"Close. The greater the instability,
        the more sensitive the system is
        to even small changes,
                which can greatly influence the path
                that the system ultimately follows.
A small number of people can,
        at that point, influence
                the entire system."

                "So even though I feel like I can't do much to
                change the world, especially when it's going
                nuts, you're saying that the opposite is true?"

"It's very possible.
        As you pointed out,
        any minority of individuals,
        engaged in passionate pursuits
                of Planetary Mind,

could bring forth a
transformation of the whole."

"But that also means some minority could
seize power and actually do even more
damage, like what happened in Europe
after World War I, when a small number of
fascists took over Germany."

"Oh, yes . . . and stories
do not have to be based on facts;
tyrants will create stories that deceive
and lead to catastrophe,
stories that persuade people to willingly
carry out destructive, even horrendous, acts.
And the greater danger is that the rest of us
will fail to stand up and stop the horror,
whether out of fear, indifference,
helplessness, or denial.
All the more reason for us
to engage in the challenge
of the Great Transformation, yes?
No matter what you're doing right now—
what *any* of us is doing right now,
this moment, consciously or not—
it is influencing the greater story
in positive ways or negative ways."

"And you're saying my influence might be
bigger, because things are so chaotic."

"It is a possibility.
You won't know until you try,
and even then, you may *never* know."

"Okay, so what am I supposed to *do?*" asked
the Seeker.

The Sage paused.

"Perhaps it would be helpful
    to look at the larger picture first,
        to use the crises of
    the Great Transformation as our guides.
    We talked about these the last time we met;
        do you recall them?"

                            "I certainly remember the one about the
                            environment."

"Yes, environment and ecology.
    In terms of these, our tasks are
        to learn to live within
        the means of Earth's bounty,
        which she provides through
        the Web of Life,
    to help the Web of Life
        heal itself where we can,
    and for us to create our artifacts
        and our cultures in ways that align
        with the essences and lessons
            of the Web of Life."

                            "That's a lot."

The Sage smiled.
    "Yes, plenty. But there are
    two more."

                            "Let's see . . . something to do with
                            spirituality, right?"

"Yes. Our tasks here are
    to reconnect with the Infinite,
        seeing ourselves as an integral
        part of the Infinite,
    to cultivate a sense of awe
        for the Universe,
    to hold the Web of Life as sacred,
        as it is indeed that which brought us forth

and sustains us now.
And the last one?"

           The Seeker thought for a moment. "I don't
           remember that one," he said, finally.

"Not surprising—we tend to not
    recognize the connections among
        ecology, spirituality,
        and our society's institutions.
    But these are critical:
        to seek greater fairness and justice,
        to learn the rewards of greater cooperation
            as we move beyond our self-interests,
        and to do the hard work of widening
            our circles of compassion,
        realizing that when we focus our
        compassion too closely
        on one part of the whole,
        we are in danger of damaging
        the whole, or other parts of the whole,
without even realizing it."

           "For example?"

"For instance, Modern Mind's
    medical technology has become
        nothing short of miraculous.
    But leaving behind the question
        of how this technology has also
        obscured the wisdom to know
        when it's time to say goodbye
        and let death take its natural course,
    it has become fantastically expensive,
    creating an uncomfortable question:
        who gets to benefit from
        this technology, when it is clear
        that not all can have access?"

"But I have to say, if you're talking about
your spouse or your kid, you're going to

want all of that technology to help them. *I*
sure would."

"Of course you would,
     as would *anyone* with a heart.
     But it still begs the question:
     is it fair that one's child
          has access to all of this
          expensive technology,
     while millions of other children
          die every day from diseases
     that are relatively inexpensive
          to prevent?

"And it begs another, even more
     disturbing paradox:
     as we save and extend the lives
          of more and more people,
     we put a greater strain
          on the ability of the Web of Life
     to provide for those people—
     and those of us in the First World
     benefit the most from this technology
          but also take the most
          from the Web of Life.
     As we provide medical care more equitably,
          what are we doing
          to reduce the pressure of
          not just human populations,
               but the amount of damage
          that human populations cause
     as they become wealthier
          on a planet that
     simply has limitations?"

"I don't get it. You're not saying we *shouldn't*
try to have better medicine, are you?"

"Of course not.
    But as we try to save every child—
        which I'm not at all against;
        it's certainly the compassionate
            thing to do—
    we must also reduce
        the total number of children born
            so that all can share
            the bounty of a healthy Web of Life.
    This is *also* the compassionate
        thing to do, lest a runaway
    population stress the Web of Life
        to a collapse,
    causing a horrifying die-off
        and unimaginable suffering
        that vastly outstrips whatever
        gains we make in our
        smaller acts of compassion.

"This larger compassion is not built in to us.
    We see a hungry person and can feed her.
    But seeing all of the ecological devastation
        caused by the collective,
        especially from the First World?
    That's not possible
        through our immediate senses.
    We need to think about and change
        the way things are set up,
        so that we encompass
            this broader view
            of compassion."

"We have to think bigger, in a way," said the
Seeker.

"And *act* bigger, too.
    Now, looking again at all our crises,
    it's vital to see that they are related.
    Injustice is related
        to ecological damage;
        both are related
        to spiritual starvation.
    Planetary Mind seeks answers
    that address the root causes
        of all three."

        "Wow." The Seeker shook his head. "That's
        a pretty tall order. You think we'll be able to
        make all that happen?"

"We don't have much choice
    but to try;
    the realities of the crises
    driving the Great Transformation
        are seeing to that—
and if we continue to ignore these realities,
the Web of Life will itself have no choice
but to lay our civilizations to rest,
    as it has done with other civilizations
    that failed to awaken and act.
But rather than focus
    only on *that* reality,
it also helps to focus on
    how we're aligning with
        the creative drive of the Infinite
    and what we might create together.
We're going to *have* to change; why not
    have a little fun in the process,
    and see what we can create?"

        "Okay, but still . . . what am I supposed to
        do?" asked the Seeker.

"That is for you to discover.
        Your task is to uncover your purpose
        as you look squarely at these realities—
                a purpose that sings to you,
                        however quietly,
                pointing to that
                which is yours to do
                        as we create Planetary Mind.
                Follow it.
                Perhaps it is small,
                perhaps it is big."

The Sage looked past the Seeker, thinking.
She watched the trees on the other side of
the playground as they swayed in the breeze.
She then looked back at him.

                        "What?" asked the Seeker, puzzled.

The Sage thought
        for another moment.

"Perhaps your work is
        to find a place on Earth,
                even a small one,
        that is in need of care,
                and take care of it,
                        learning what you can from it,
        and teaching others
                what you have learned."

                        The Seeker pondered her words.

The Sage said nothing for few moments.
Then she drew in a breath, let it out, and
shrugged her shoulders.

"The thing is, *you* are the one
                who must decide
        what your contribution will be,

however you uncover
where your passions meet
   the world's needs.
Each person will create her life's dance
   as we collectively face
the Great Transformation,
creating something never seen before."

> "But how do I decide? And if this thing we're
> creating hasn't ever been seen before, if we've
> never faced anything like this, how can we
> expect to know what in the world we're
> doing?"

"Excellent questions,"
   said the Sage.

WISDOM AND AUTHORITY

They watched as the man called to the
children playing. The children ran over to
the man, who cleaned the sand off of their
hands and then gave them crackers, apple
slices, and carrot sticks to eat.

The Sage nodded toward them.

"Just as these two children have
    little idea what lies before them
    as they grow into adolescence
    and face the challenges
       of adulthood,
    they will still be able to call on
       untapped reserves and intuition,
      as well as their learning.
    So too it may be for us
      as we create the second path
    through the Great Transformation."

        "We have to figure it out as we go?" asked
        the Seeker.

"Yes, but remember, the Web of Life
    has faced catastrophe before,
    and the Infinite's process
       has Its own wisdom
      beyond logical comprehension.
The authority we seek will
come from within us
    individually and collectively
as we tap into this inner wisdom,
    to help us find our way."

        "That sounds so vague. How do you tap into
        this?"

"Any number of ways:
     seeing and challenging the
          unexamined assumptions
          of Modern Mind's worldview,
     learning to harness
          the wisdom of the collective,
     tricking the mind to get
          out of the way,
     paying attention to your body
          and its sensations.

"That anecdote you told me,
     about when you were a child
          crying in your father's arms,
     reveals a deep passion and caring.
     It is connected to your inner wisdom;
     it will help guide you if you keep it close
          and don't bury it.

"This passion is connected with no less than
     the creative force of the Infinite,
          which created the cosmos, Earth, and us.
     That connection, I believe,
     is reawakening within humanity now."

                         "I get the passion part, but it still sounds
                         vague."

"Let me use a little trick I recently
     learned as one example.
     It's from a teacher
          of sustainable architecture,
     who uses a question
     to jar his students out of their
     habitual thinking.

"When they're confronted with
     an architectural challenge,
          he finds that, left to themselves,

with their thinking steeped
in the worldview of Modern Mind,
their solutions tend to focus
        on technological fixes
        to sustainability.

"But when they answer just one
        question, they tend to create designs
                that are strikingly full of life:
        'Of the options available, which looks
        more like a picture of your own soul?'"

                          "Huh. A picture of your own soul," mused
                          the Seeker.

"This question does a delightful job
        of bypassing all the static
                of Modern Mind's
        analysis and logic."

                          "Give me an example."

"Let's use one from your work.
        When you take into account
        your usual criteria
                when building a house,
        which should you choose—
        one that's built
                as inexpensively as possible,
        or one that's built
                as carefully as possible?"

                          "Well, you *always* want to build as carefully
                          as possible . . . although sometimes budgets
                          cause you to cut corners. There's a lot you
                          have to consider: how much money you
                          have to work with, the contract on the land
                          and when it has to close, cost of labor, cost
                          of materials."

"Now, let's reframe
the question:
'Which looks more like a picture
of your own soul:
a house that's built
as inexpensively as possible,
or a house that's built
as carefully as possible?'"

The Seeker laughed, and raised his eyebrows.
"Wow, yeah . . . the answer leaps out, at least
in my mind—the one built as carefully as
possible. Every time."

"Isn't that a wonderful question?
It magically takes you out
of your habitual, narrow views
and taps into something
that *knows* what to do.
You don't even have
to believe in the soul."

"Yeah, but we live in the real world, and
sometimes—a lot of times—there just isn't
the money to do it the careful way." He
pondered the idea. "Although that sounds
kind of crazy, doesn't it."

"It certainly does—you've bumped up
against the way things are set up.
Questions like these
help to reveal the craziness
in our systems.
They also show us how to get to the
inner wisdom that
Modern Mind has walled off,
so to speak.

"But this example shows a way of
       getting out of habitual thinking
       on an *individual* level.
    Let's look at building a house
       from the perspective of harnessing
       the wisdom of the collective.

"If you asked all the people involved
    in building the house to a meeting,
       including the developer,
       the city planners,
       the bankers,
       the builders,
       the neighbors,
   and the ultimate buyers,
       they'd all come with different ideas
       of how the house should be built.
    Someone unskilled
       in handling such meetings
       can quickly create gridlock
       and bad feelings."

                    "Oh, yeah, I've sure been *there*." The Seeker
                    shook his head. "Those meetings are train
                    wrecks."

"But if you brought in someone
    who *was* skilled,
    who helped these stakeholders
       deliberate the issues
       and information,
    indeed, made sure to bring
       good information from all sources,
    made sure that there *were*
       people with different backgrounds
       and ideas in the room,
    and made sure the group
       fully heard each person,

it's very possible that the group
could find common ground,
even among people with
disparate positions.
Done well, the worst that could
happen is that the members
will at least have more trust
and respect for one another.
The best that can happen is
that a strong collective authority
emerges from the group."

"Hmm," said the Seeker. "But I don't see
how that changes the way things are set
up. You still have to go through planning
commissions and city councils, and they're
just not set up that way."

"Ah, but if you hold such
meetings *outside* of that system,
making certain the meetings are
as transparent and open as possible,
the stakeholders would likely show up
to these commissions and councils
with a much clearer collective authority.

"Tell me, what might occur
if this were to happen several times
in a particular town?"

"I could see it helping, yes. But not changing
the way things are set up. You still legally
have to have the councils and commissions."

"Yet if there were enough
committed people who helped
keep these meetings going
for every major issue in a town,
such meetings could become

a part of the way things are set up
        for this town."

                                "Maybe," said the Seeker. "So you don't
                                lose the legal part of doing things, you're
                                supplementing the process."

"Exactly, and that's not
        out of the question.
    But let's return to your
            earlier observation,
    how crazy it sounded that
    there sometimes isn't money
            to do things the careful way.
    What would have to be true
            for there *not* to be a dichotomy
            between doing things
            the careful way versus
            the inexpensive way?"

                                "Oh, man, I don't know that that could
                                happen." The Seeker thought. "Well, for
                                one thing, you'd have to have more than just
                                money be the bottom line. I suppose that's
                                why we have things like building codes to
                                keep us to a certain standard. But beyond
                                that—that question's a stretch."

"How about there being an
        agreement that building with care,
        building a quality home,
        is important for its own sake?"

                                "Now that's a *real* stretch!"

"Yet go to the older part of
        some of our towns and cities,
            and you'll see that agreement
            built into the gorgeous homes

we used to build
as a matter of course."

The Seeker raised his eyebrow. "Yeah, well,
things were cheaper then . . ." He paused,
thinking. "But you've got a point. Though I
don't know how we'd get back to that kind
of attitude."

"These are questions worth asking
    and tasks worth taking on—
    examining what would need to happen
        to change the way things are set up
        and challenging the stories underneath
            those systems.

"These are examples of
    the many things we can try
    as we collectively and individually
    explore those untapped reserves
    and intuitions I spoke of earlier."

"Even though we're bound to make mistakes
    as we go along."

"Of course, and that's nothing
    to be afraid of, because that's
    the way the Web of Life has always done it:
    through endless creative experimentation:
        thousands of ideas,
        thousands of experiments.

"When we mindfully plant a garden,
    we notice where the Sun shines
        and the water flows.
    We plant our seeds, then watch,
        encouraging those plants that thrive,
        helping or replanting
        those that struggle,
    all the while cultivating

a sense of deep humility
as we learn from the garden
    and teach its lessons
        to one another.

"We will sow many seeds,
        experimenting with new ways of being
            within the Web of Life,
            and with each other.
Failure will usually not be a problem,
        as long as we learn the lessons
        that lie within our failures."

The man at the playground started packing
up the snacks, his book, and the children's
belongings. He called to them, telling them
it was time to go.

The Sage rose from the bench.
    "Let these thoughts roll around
    in you awhile. Play with them,
    experiment with them,
    and we'll see what happens next."

                    "Just like the Web of Life, right?" asked the
                    Seeker, also rising.

The Sage laughed delightedly.
    "Yes! Just like that!"

SEEDS

Seeds scatter all around us.

To be a seed is simply to be.

It is not to achieve,
    although achievements may happen.
It is not to progress,
    although progress may happen.

It is the nature of the Web of Life,
    in all its creations,
    to scatter vastly more seeds
        than will ever come to fruit—
each containing the coiled codes of life
    billions of years old,
each containing the potential
    for an entire being,
each searching for the niche
    that furthers the story of
        the Web of Life.

The apple tree creates hundreds of apples
    thousands of seeds
        within the lifetime of the tree;
        only a few seeds will become trees.

This, then, is a design of the Infinite:
    to cast forth many seeds
        that they may seek perfect possibilities
        and find small slivers of sustenance
    that shall allow those few to flourish.

For the success of the few
    rides on the shoulders of
        the many
        that never become more than seeds,

that need never become
more than seeds.

So, then:

The Universe seeds itself
with billions of galaxies.
The galaxies seed themselves
with billions of stars.
The stars seed themselves
with billions of planets.

And on Earth:

billions of seeds
for each species:
millions of species
seeding the Web of Life:

and thousands of ideas
for each human achievement.

A teacher once said,
"Many are called, but few are chosen."

Yet the few that are chosen
need the many that are called.
For with too few seeds, the one seed
may never find its place and take hold—
and the species fades away.

There must be thousands that do not survive
for the one to thrive.
This is the paradox of creativity,
and the purpose
of those that do not survive.

What we see as the failures of the thousands—
false starts,
dead ends,
and mistakes—

are transformed into
the exalting joy of all.

  And all that do not survive
    share in the majesty
    of the one that thrives.

STORIES OF MODERN MIND

Workers moved back and forth, filling the large blue debris bin on the edge of the job site, piling on wood scraps, plastic piping, broken glass, and old plumbing fixtures.

The Seeker stood by piles of paint cans that sat next to the bin, which he planned to take to the hazardous waste dump, though he was skeptical as to what really happened to them.

"Sure doesn't look like a picture of anyone's soul," he muttered to himself. He hated this part of construction—the wood and fixtures being tossed away were often perfectly useable. And wasn't there something wrong about throwing all this stuff into a landfill? Something at odds with one of the Essences of the Web of Life, that part about flow?

"When's the truck coming for pickup?" asked one of the workers. "We'll definitely have this filled up by tonight."

"It was scheduled to be here tomorrow afternoon, but I'll call them and see if they can get here first thing in the morning."

"That'd be good. Probably have at least one more load."

The Seeker nodded, silently clenching his teeth.

The worker stopped and put his hands on his hips as he looked at the pile. Then he shook his head. "Seems like such a waste, doesn't it, to toss all of this stuff? Lot of it's still good."

The Seeker looked at the worker, startled. "Yeah," he stammered. "Yeah. Too bad we can't figure out how to reuse it."

"Ah, well. Maybe someday we'll figure it out," said the worker, shrugging his shoulders and turning back for another load.

The Seeker watched after him, not quite sure what to think.

◆

"Hah, you see?"
        exclaimed the Sage.
    "Not everyone is
        sleepwalking around!"

"It surprised me. I'm not sure what exactly it means, or what to do about it, but at least it's something," said the Seeker.

They were sitting at the table where they had first met, in the small patio behind the coffee shop.

The owner had earlier watered the two dozen or so potted plants that she kept on the patio, and the air was humid as the water drained onto the terra cotta tiles. The patio was shaded by a trellis, through which a wisteria vine had twined itself. Bees

hummed above, moving among the lavender
blossoms.

"And you're right; there's definitely
     a problem with how we
     'throw things away,'"
        said the Sage.
    "Yet another example of bumping up
       against how things are set up.
      But tell me, what bothers
         you most about it?"

                     "The fact that it's not being used well, that
                     all that wood and stuff is just being tossed."

"Why isn't it wasteful
    when an old tree dies, falls,
    and rots in the forest?"

                     The Seeker sighed and thought a moment.
                     "It's . . . well . . . it's not out of place; isn't it
                     part of the process, dying and rotting?"

                     He leaned forward. "Cutting it down
                     thoughtlessly and tossing a lot of it aside . . .
                     that just seems so heartless and ugly."

"'Part of the process'
    is precisely on target.
       Rotting is indeed
   an integral part of
      the Essence of Flow.
Wood tossed in a landfill
      doesn't rot properly,
      doesn't flow through
      the Web of Life.
The ugliness and heartlessness
      of the landfill
betrays its underlying dysfunction.

"But let's also look at the term 'away.'
   What do we mean by this?"

                              "Someplace else," said the Seeker.
                              "Somewhere we just put our junk out of
                                 sight, bury it, ignore it or hope someone else
                                 takes care of it."

"Out of sight, but not out of flow.
   Even though our landfills
      interfere with flow,
      by no means are they
      outside of flow.
   Why might that be a problem?"

                              "We have to be careful what we put into
                                 them."

"Yes, to put it mildly.
   Modern Mind fools itself
   into thinking that it could
   simply take its wastes,
   especially dangerous ones,
      and bury them or toss them
      into rivers and oceans.
   But when it does so,
      these wastes enter into flow,
      in spite of our efforts
      to contain them.

"So, on the one hand, some of the ways
      we deal with wastes, such as landfills,
      are dysfunctional and interfere with flow.
   On the other hand, there really is
      no way to sequester dangerous wastes
      without tremendous expense,
      and even then, there can be
      no guarantee that such sequestration
         will work over time.

Sooner or later,
    the Web of Life will likely make
    short order of our attempts
    to bury our wastes.
And in any case, if these wastes are
    materials that the Web of Life
    hasn't seen before,
they will throw it into dysfunction.

"It's sobering, when you
    step into a large store,
        look at all the merchandise,
        and realize that nearly all of it will,
            quite soon,
        end up in a landfill,
    or scattered across the landscape."

                "And I *hate* that . . . but I'm just as guilty. I
                mean, can you really live in our society and
                not cause damage? Can't we just wake up
                and start creating laws so we don't *do* all of
                this insane damage?"

"But can you really change things
    merely by changing laws?"

                "Well, *what*, then? I just . . . I keep coming
                back to that. How else do we make such
                huge changes? We can't just keep doing this!"

"By doing more than changing the rules.
    Yes, laws are important,
    but many countries—
        the United States and China, for instance—
    have strong environmental laws,
        but they aren't always heeded,
            to say the least.

"We need to approach changing
    the way we've set things up

from this angle *and* many others
as well, such as
    the way information flows
        and how transparently it flows,
    the rules of the system itself,
    the goals of the system—
and underneath all of these is
    the worldview,
    the stories
        that, consciously or not,
    permeate and inform
    the way we've set things up.

"The problem is that
    it's easy to change the players
        in our halls of power.
    Though it is a more difficult, messy,
        chaotic, and organic process,
        it is far more effective
        to change not just rules, laws,
            goals, and regulations
        but also the stories that run
        like a deep, invisible current
            beneath those halls.

"Perhaps it would be helpful
    if we take a closer look
    at some more of these stories?"

                "Yes, please," replied the Seeker.

PRIVILEGE

The Sage took in a breath, and let it out as
she looked up and watched the bees as they
worked from blossom to blossom.

"Let's start with privilege, specifically
        as it relates to a person's
            ethnicity and gender.
    The story, at least in our culture,
    is that, fundamentally, racism and sexism
            are pretty much over;
    yes, a lot of problems linger
            that we need to attend to,
    but deep down, there's a story
    that, even though there are racist
            and sexist people,
    we are not a racist or sexist culture.

"Yet this is far from the case.
    Racism and sexism are still alive
            and everywhere in our culture,
            are *endemic* to our culture."

                        The Seeker shifted in his chair.

"Not the most comfortable
    of topics, yes?"

                        The Seeker gave a nervous laugh and
                        nodded. "Yeah, that's for sure." He thought
                        for a moment. "I guess I see the sexist part,
                        but not so much of the racist part. I mean,
                        yeah, there's still a lot of racism, I guess, but
                        I just don't see how it's that pervasive in the
                        whole system."

"Largely because you don't
    have to see it—

that's part of your privilege,
>not having to see
what doesn't immediately affect you.
But just start with looking, really *looking*
>at advertisements,
>television shows,
>movies,
>and pictures of CEOs
from the perspective of someone
>who isn't white or male.
Look at the evidence around you
and listen—don't just *talk*—
>to women and people of color.
>You'll start to see it
>everywhere."

"Really."

"Yes, really."

The Sage thought for a moment.
>"Let's do this—think back
to a time in your life
>when you felt judged,
>when you felt like you didn't fit in,
because of your appearance
or your background."

The Seeker looked at his coffee cup in front
of him, thinking. "Yeah, okay, there was
this one time, when I was dating a girl who
was in a private school way back when. I
remember going to a party with her, and
how out of place I felt. The way the other
kids were dressed, the way they talked about
their vacations and cars and everything . . .
I mean, my folks were not really wealthy,
but it wasn't like I was dressed in rags or
anything.

"Even so, I really felt out of my element. At one point, I overheard one kid say to another, 'Dude, let's blow off school tomorrow and head out to my parents' beach house.'"

The Seeker shook his head. "I remember thinking something like, 'Man, this place sure ain't *my* place.'

"The weird thing is, nobody said anything, nobody asked me to leave or was mean or anything, but it was just something in the air, something unsaid, that felt like '*You don't belong.*'"

The Sage nodded.
    "It must have been
    quite painful, yes?"

"It was a long time ago, but . . ." He stopped and looked away, remembering. Then he looked down. "Yeah, it was a little painful. I sometimes wonder if that's why we broke up, not over that particular night, but because I was just . . . different, I guess."

The Sage sat quietly.

"Most everyone alive has had
    similar experiences,"
    she continued after a moment.

She sighed.
    "Even this morning,
    a white woman came up
        and stood next to me
        as I was waiting to be served
        at a delicatessen.
    The server behind the counter

finished with the customer
in front of me and said,
'Who's next?' He looked at me,
then looked at her and raised his
eyebrows and tilted his head up
toward her, expecting her order.
It happened so fast, I'm *sure*
he was not conscious of it.
To the woman's credit, she motioned
toward me and said, 'She's next.'

"So as he takes my order and prepares the food,
I silently ask 'the questions':

What just happened?
What's going on here?

Am I just seeing things?
Should I say something?
What should I say?
Do I really want to deal
with whatever reaction
he'll come up with?

Am I being overly sensitive?
Did he make an honest mistake?
But is that mistake ever honest,
if it truly was something
that was based on
the color of my skin?

"On and on. I can't help but ask
these question whenever these
micro-moments happen."

"Good God . . . I had no idea," said the
Seeker. He looked down at his hands,
absent-mindedly rubbing his thumb against
his forefinger. "How do you let that not . . .
how do you deal with it?"

She smiled and shook her head.
  She picked up her tea,
  took a sip, and sighed
      as she set it down.
  "Sometimes it's not easy.
  If I say nothing as it is happening,
      I get to deal with my anger
      for staying quiet;
  if I *do* say something,
  I get to deal with whatever
  spinout might happen.
      Either option is challenging.
  But whatever I do, well . . . I breathe,
      try to not let it continue to wound,
      and try to see the humanity
          of the person.

"And here's something really important:
  remember that feeling of
      'You don't belong'?
  Notice that you had to reach back
      a number of years to find
      a time when you felt this.
  A lot of people of color—
      and women, for that matter—
      have that feeling *every day*,
  as they face small and large indignities
      every day.

"*That's* the difference," she said quietly.
  "*That's* what people
      who enjoy privilege
      really need to know
  if they want to help change things."

The two of them sat silently, listening to the
hum of the bees above.

"I don't know what to say," said the Seeker, finally.

"There's nothing you need
        to say right now.
    What's important to me
    is that you're simply listening."

"Okay. I can at least do that."

The Sage leaned back, sighed,
    and smiled again.
    "Thank you," she said.

"You're welcome."

They sat together quietly.

After a few moments, the Seeker said, "Um, I have a couple of questions, but . . . well, I guess I'm afraid of saying something stupid."

The Sage laughed.
    "Oh, at some point,
    you probably will,
    but don't make a big deal
        out of it if you do.
    Just apologize and move on."

The Seeker let out a breath. "Okay, here's the first: back in the story of Ancient Mind, you said some clans looked down on people in other clans, that they saw them as less than human. So this is more than a matter of Modern Mind's story, isn't it?"

"Oh, yes, all human cultures
    are subject to this shadow side,
    of how we create stories
        about the 'other.'

"However, Modern Mind
         put a new twist
    on seeing and dealing with the other:
    it institutionalized its prejudices
         on a large scale,
    canonized them in its laws and societies.

"Now, as we struggle
    to rid ourselves of these structures,
    we still face insidious, unintentional,
    and unexamined racism
         that continues in spite of our efforts,
    and this is as big an issue
         as overt bigotry, sexism or snobbery.
    People often don't *mean*
         to behave this way,
         yet it's still lodged
         deep within our institutions
    as well as ourselves."

                    "So even though the written law changed,
                    our behavior hasn't."

"Not fully, nor have the *systemic*
    behaviors fully changed."

                    "Systemic, meaning what?"

"For example, there's a private school
    where a friend of mine works.
    She and the staff are kind
         and wonderful people.
    Nobody in this school
         of primarily white kids
         overtly sets out to exclude
              minority children,
         yet it still happens,
              even with the best of intentions.
    No master plan of any city

overtly looks for poor neighborhoods
to put polluting industries next to,
yet it still happens,
    even with the best of intentions.
No one sets out to deprive
those same poor neighborhoods
of stores selling fresh food,
yet it still happens,
    even with the best of intentions.
And so on.
    Part of the work we need to do
        includes changing
these larger meta-structures
that create these kinds
of systemic problems."

"Well, that leads to my next question that
I'm kind of afraid to ask. I mean, when I
think about the Sixth Extinction . . . I don't
know . . . I think I'm getting what you're
saying, but how big is the issue of privilege
compared to the collapse of the Web of Life?
To me, that sort of trumps everything."

He shifted in his chair. "Although I suppose
if I were poor and lived next to a power
plant, I'd see it differently."

"You most certainly would.
    This again reflects the essential
        sense of separation
        in the story of Modern Mind:
            that these issues are unrelated.
But as we've seen before
it's not that addressing issues
    of social justice
    are more or less critical
    than ecological sustainability;

they're the same.
They're not just related:
they're *inseparable.*
In much the same way that
Modern Mind sees
a planet full of 'resources'
that can be used and thrown away,
it sees people that can be used
and thrown away.

"Working on issues of privilege
is *not* a distraction or a side road—
this work instead cuts deeply into
the same story of separation
that is at the core of Modern Mind's
assault on the Web of Life."

"My God." The Seeker sat back in his chair
and stared past the fence into the street. "I
never thought of it that way."

He sat silently, thinking. "Okay, so how do
you go about working on privilege?"

"You and I are doing it—or rather,
we've *started* it.
Simply naming it, talking about it,
acknowledging that privilege exists
is a big step.
But it *is* only the first step,
and there's more
that I can suggest for you to do.
But remember that it *does* fall on you
to do this work.
The 'problem' of privilege
is as much yours to confront
as it is mine."

The Seeker pondered. "I guess . . . Well, I'm
still afraid that I'll say or do something that
just confirms that . . . that I'm racist, or
whatever."

"Well, you can't have been raised
    in this culture
and not have such proclivities,
so you can let go of *that* worry.
The work is *not* about
    determining if you're racist
    or sexist or classist,
but is instead about recognizing
    these proclivities in yourself
    and changing how you respond.
But more important,
    your fear could hold you back
    from doing this work—
    and part of your privilege
    is that you *can* choose to ignore it,
        simply because incidents
        like the one in the delicatessen
        seldom happen to you.
You don't *have* to confront
this reality within you,
within our culture.

"This raises the question, then;
    if you don't *have* to do this work,
        why do it?
Why explore this uncharted
and uncomfortable territory?"

The Seeker looked down, thinking. "Well,
like you just said, it's related to the
ecological crisis. You deal with one, and
you're dealing with the other, too. And
I guess . . . it's just the right thing to do,

especially if we want to change, to help
people who have to deal with being put
down day in and day out."

"Yes, and it's more than just
    the right thing to do
    for people of color or poor people
            or women—
        for your wife and daughter,
            for that matter.
The rewards are potentially enormous
            for all people,
        even beyond changing
        the way things are set up.
Just as rewarding is this:
        as we engage in these dialogues,
        we see past the identities we've been given
            and that we've taken on;
            we see through them to
            our common humanity.
Amazing connections between people
        can happen, in spite of
        the atrocities of the past.
        Maybe it'll take a long time
            to fully heal those wounds,
            if they ever can be healed,
        but we *can* begin to see through them,
        see one another more deeply
            right now.

"Even this small conversation
        has allowed us
        to see one another
    a little more clearly, yes?"

                    The Seeker smiled and nodded. "Yes, it has."

"In a sense, *your* liberation
        is tied in with this work, too.

Privilege binds not only
    those who are put down,
but also those who are privileged.

"As we change our story
    to acknowledge
that these issues of class, gender, and race
    do exist in a systematic way—
that they've been instilled into
    our neurology by our cultures,
that they're things that people
    who enjoy privilege
    need to work on as much
    as everyone else, if not more—
that will help to bring the spirit of our laws
    more fully into fruition
        as we address these issues.

"The story of Planetary Mind
    includes this:
we have work to do—
learning how to talk about
    racism, sexism, and classism
        in ways that heal
        and bring us together
    in the common humanity
    that lies underneath
        our superficial identities.
But it is powerful work,
    both for those of us with privilege
    and those without.

"And there are even greater rewards.
    When people passionate
        for their smaller causes
    see past their comfortable boundaries
    and begin to bridge these
        painful divides,

to the point where they realize
        they're all working
on the same core crises,
        what happens then?"

                    "Huh. Wow. So instead of fighting one
                    another, they can join together. They're
                    stronger."

"It is a momentous day indeed
        when people let go
of their prejudices, suspicions,
            habits, and hatreds
and discover the astonishing power
                that is forged
        when they meet one another
as humans who are connected,
        rather than identities that are separate.

"As we bridge these divides,
        we pave the way
            for greater cooperation and resilience
        and also discover just how much power
                we can have together.

"Let's next look at how
        changing this story of separation
        changes another story
        we have about the Web of Life."

A SENSE OF SACREDNESS

A hummingbird arrived and sipped at the
trumpet creeper growing alongside the
patio's fence. It flew up to a tree just outside
the fence of the patio and settled on one of
the branches. After a moment, it flew off.

"As the story goes," said the Sage,
    "Modern Mind sees itself
    as fundamentally separate
        from the Universe.
We've seen what happens
as Modern Mind's story of separation
        inflames the sense
        of 'otherness' we have.
What attitude does
    this sense of separation
    create in Modern Mind
    regarding the material world?"

                "Attitude . . . well, that we're separate from it,
                  and it's all ours, that we can do whatever we
                  want with it," said the Seeker.

"Everything on and in Earth—
    even on the Moon, asteroids
    and other planets,
        if we can get to them—
    is called a 'resource,'
        to be used
    as Modern Mind wills.

"This word 'resource' contains little sense
        of reciprocity,
        of give and take,
        of interdependence,
            which is more in line with

the Essences of the Web of Life.
It betrays an underlying sense of exploitation,
of use only for gain—
often to the detriment
of that which has the misfortune
of being labeled a 'resource.'
Even our companies have
departments called
'Human Resources,'
which, I suppose,
is a little better than 'Personnel,'
but both terms reflect
the same worldview.

"Planetary Mind takes issue
with seeing oceans, deserts, and forests
as mere 'resources,'
for it acknowledges the myriad
living things in each of
these ecosystems—
birds, animals, fish, trees, insects,
and even people:
all of these are expressions
of the Web of Life.
Planetary Mind recognizes our complete
dependence on—our *interdependence* with—
the Web of Life.
Planetary Mind would no more use the word
'resource' to describe a tree
than it would call
one's mother a resource.
Modern Mind over-consumes
because it *uses* words
like 'consume' and 'resource'
in reference to the Web of Life,
words that belie Modern Mind's
fundamental belief in separateness."

"What's a better word than 'resource?'"

"For me, it's the word 'bounty.'
   The Web of Life, Earth, and
   all the Universe are sacred,
      worthy of respect,
      worthy of our reverence
        and our awe.
   The Web of Life,
      which offers its bounty to us
        that we might live,
     is deserving of our
   deepest respect and gratitude."

"But that's . . . how does that work? Are you
saying that we can't cut down trees to build
houses?" asked the Seeker.

"Not at all, nor does it mean
   that we cannot harvest food,
      farm the soil,
      or fish the streams and oceans.
   As one sage put it,
      to live, we must daily break the body
      and shed the blood of the Web of Life.

"What it means is *this:*
   when we cut down trees,
      fish the waters,
      break the soil,
   and kill for food,
   we do so reverently and with gratitude—
      as well as intelligently and mindfully,
      in ways that are in harmony with
        the lessons of the Web of Life,
      in ways that help the Web of Life
        replenish itself.

"The Great Transformation
    does not give us a technological challenge
        as much as it gives us
    a spiritual challenge:
        the challenge of acknowledging Earth's aliveness,
        of seeing the Web of Life as sacred,
        and embracing this sense of sacredness
            as one of our deepest values.

"At the same time it gives us
    a social challenge:
        the challenge of learning
        how to more skillfully
            gather wisdom from the collective
        and heal the wounds that divide us
            across classes and ethnicities,
    and set things up so that it's easy
        to act in ways that are in line
        with the lessons of the Web of Life.

"Planetary Mind, therefore,
        holds the Web of Life as sacred.
    It understands that its bounty is not
        to be mindlessly manipulated
        or exploited purely for profit
            or for personal edification or amusement.
    It respects the sacred scrolls of life
        inscribed by the hand of the Infinite
        over billions of years.
    It understands that as we align ourselves
        and our societies
        with the intelligence of Earth,
            she will respond.
    It reveres the Web of Life as a teacher,
        and learns from it."

                "A teacher . . . I think I get it," said the
                Seeker. "Planetary Mind respects what

the Web of Life has created, as opposed
to Modern Mind, which does whatever it
wants to the Web of Life, because it sees it as
fundamentally stupid or unintelligent."

"Yes, and if we study
    the Web of Life and its essences
    with a sense of respect,
        reverence, and awe,
    it will teach us how we might survive
            and thrive through
    the Great Transformation.
We began today talking about
    the Essence of Flow;
    remember the other essences
    of the Web of Life?"

                    "Besides flow . . . I can remember a couple of
                    them: flexibility and relationships?"

"Relation, actually, and also
        emergence and unpredictability,
        beauty,
        and creation and destruction.
    Each essence holds lessons.
    Becoming students of these essences
    is one way to learn how we might recreate
        our cultures and societies
        and how we might create
        the comforts and necessities
            for our lives.

"These essences will *always* trump
    any economic 'laws' we invent.
    If our human societies are
        to survive the Great Transformation,
        they must align their worldviews
        and actions so as to be in
        harmony with these lessons."

"For example?"

"Okay . . . let's look at
        the Essence of Relation,
        using you as an example.
        Where does your body begin and end?
        Where might most people say that
                'you' end?"

                                "That I begin and end with my skin," replied
                                the Seeker.

"Agreed. If I drew
        a detailed picture of your body
                on a piece of paper,
        most would say this
        represents your body,
                whole and complete.

"Your answer reflects
        the deeper story of Modern Mind
                that we've been talking about,
                        that sees everything as separate.
        But when you look
        with a broader awareness,
        larger pieces of the picture emerge.
        How could you exist,
                whole and complete,
                        with just your body
                        and nothing else?
        How might you breathe,
                for instance?"

                                "I see; the picture doesn't fully show what I
                                need to be alive."

"Indeed; to be more complete,
        it must also show the trees
        that you depend upon
        for oxygen, yes?"

"I think I see where you're going. Because
now I have to do the same thing with trees:
a picture of a tree must also show what the
tree needs."

"Exactly . . . yes, such a picture
must also include
you, and other animals
providing air for the tree,
clouds that bring rain,
decaying plants and animals
commingling with broken rock
to create the soil
in which it grows,
microbes in the soil
that help make nutrients
available to the tree,
as well as the Sun,
which provides starlight
energy for the tree.

"All this forms a vast network
that is the Web of Life,
with no one species at the center."

"So not only would the picture of me be
incomplete without the tree and all it
depends upon," mused the Seeker, "but also
everything else in addition to the tree that I
depend upon."

"You see how impossible the task is?
You cannot paint a *complete* picture
of any organism;
to do so would necessarily include
all the organisms *it* depends upon,
as well the other organisms
these organisms depend upon in turn.

That doesn't mean you shouldn't
    paint such pictures
    or study individual species,
        as long as you remain aware
        that any sense of separateness
    is illusory.

"What, then, might we learn
    from the Essence of Relation,
    as it applies to our societies?"

        "Hmm . . . that everything's related to
        everything else?" said the Seeker, thinking
        aloud. "If one thing gets damaged, that
        would ripple through the whole thing, and
        you don't know where it might come back
        and bite you."

The Sage laughed.
    "Not bad . . . but give me
    a specific example."

        "Okay." The Seeker frowned for a moment,
        then his face lit up. "I know . . . I remember
        hearing that some of the damage that's being
        caused by hurricanes is now worse, because
        there aren't as many acres of wetlands to
        absorb the storm surges."

"Excellent," said the Sage, smiling.
    "I like this example,
        because it shows
        that 'relation' is not
        just limited to living species,
    but to the larger aliveness of Earth.

"When we hold the Web of Life as sacred,
    it serves us in many ways:
    it guides us to how
        we might help it heal

and allows us to benefit
from its bounty;
it helps us discover
how we might share that bounty
equitably and enjoyably;
it reconnects us to the mystery
that created and sustains us.

"But when Modern Mind
organizes its societies
and economies as if the Web of Life
were *subservient* to
its societies and economies,
it's asking for big trouble.

"For while the Web of Life
can be damaged,
perhaps even destroyed,
it can never be mastered,
dominated,
or controlled.

"When Modern Mind
builds roads, buildings, and dams,
creates enormous mono-crop farms,
and otherwise mindlessly
engages in vast projects
without a sense of sacredness,
without paying attention to
how its actions reverberate
throughout the Web of Life,
it truly knows not what it does.
Just as a child plucking
at the strands of a spider's web
cannot know which strand,
when finally severed,
suddenly causes the entire web
to collapse and blow away,

so, too, we cannot know what final insult
 to the Web of Life
   might cause it to finally collapse
    and send humanity's
    civilizations into oblivion.

"Planetary Mind is clear
 on this nonnegotiable reality:
   that any society, any economy,
   must recognize its absolute
   dependence upon the Web of Life
   and respect and pay homage to it
 and create its societies to be in harmony with it,
    as it is the Web of Life
    that ultimately supports
     any society."

        "We should respect the limits of what the
        Web of Life can handle," said the Seeker.

"We have to, or, as you say,
  we will get bitten.
 Even further, we need to continue
   creating ways of seeing and understanding
   the consequences of our actions,
   and changing the way we do things
   when our actions cause trouble.
 Planetary Mind follows
   the lessons of the Web of Life,
   because it knows and understands
   that it is intimately connected with
   the Web of Life, indeed that it *is*
    the Web of Life;
   it is not separate from it,
    cannot be separate from it.
 Planetary Mind knows that whatever
   humanity does to the Web of Life,
    it does to itself.

"And there are more
    lessons we might pull from
    the Essences of the Web of Life,
    but there's another danger,
        if we act without
        a sense of sacredness.
If we make this inquiry
        into the Essences of
        the Web of Life
without a deep sense of sacredness,
without a change in our worldview,
then these lessons themselves become
        yet *another* resource,
        yet another way
        Modern Mind might exploit
        the Web of Life."

                    "I don't get it . . . how could that happen?
                    I mean, if you're studying the Web of Life,
                    wouldn't that naturally put you in the right
                    frame of mind?"

"Let's talk about that,"
    said the Sage.

REVERENT TECHNOLOGY

The hummingbird returned. A second one
appeared and tried to visit the flowers, but
the first chased after it, and the two flew off
chattering and fighting as they spiraled away
spinning around each other.

"There are those who might study
    the Essences of the Web of Life,
        to learn its secrets
    only so they might exploit its lessons,"
        said the Sage.

"For example, they might
    study the bodies of fish
    as they move through water
    to create cars that
    move more fluidly and efficiently
    through the air."

                    "What's wrong with that?" asked the Seeker.

"They make a more efficient *car*.
    But cars themselves
    are an enormous problem.
    Yes, an efficient car burns less gas,
        yet we still increase the amount
        of fuel we burn overall
        if we continue to increase
        the total number of cars on the road.
    And cars caused us to create inefficient cities
        sprawled across the landscape;
        what good is an efficient car
            in an inefficient city?
        What good is *any* device
            that does only slightly less harm?
        What good is a building

built with the latest in
'green' technology
if it houses a corporation
bent on mining fossil fuels?

"Modern Mind strives to
create technological fixes to our crises
with the same worldview
that caused the technological messes
in the first place.
It dreams of technotopia,
using technology to control,
overcome, and dominate
the Web of Life.

"Even the 'good' metaphors it comes up with
that have the intent of creating
a more sustainable world
can be dangerous."

"Such as?"

"Such as seeing ourselves
as managers of the Web of Life,
or seeing it as a machine.
There's a danger in thinking that
we can manipulate this machine
and are smarter than this machine.
There is a dearth of respect,
a lack of humility
in such an attitude.

"A more useful metaphor,
and perhaps more accurate,
views Earth as alive
in her own right,
and that we are intertwined
with her own mysterious,
innate intelligence,

with her own ways of being
    in the Universe.
Any 'management' we do—
which, incidentally, *is* important work,
and I don't mean to discount *all* such efforts—
has more to do with managing *ourselves,*
    of coming into right relationship
        with the Web of Life
and doing so with a sense of humility
    as well as respect.
The Web of Life is so deeply complex,
      it is well beyond
    our capacity to anticipate
    all the consequences of our actions.
We've seen time and again
    how quickly things go awry
    in spite of our best intentions
    when we try to fix things,
      as opposed to helping
    the Web of Life heal itself."

                "I see . . . so it takes more than studying the
                Web of Life, it also takes changing the way
                you look at it, with the sense of sacredness
                we just talked about."

"Exactly. As we awaken
    from the story of Modern Mind,
  we must see beyond
      our minds and our logic,
      beyond the needed technological fixes,
        beyond seeing
      that we have tremendously diminished
        Earth's capacity to carry life.
We must also see with
    our hearts and our intuition,
      see also that we have wounded Earth,
taking more and more from her—

more than she can continue to give—
without regard for her,
her other children,
and the grandchildren of all species,
including our own."

"I get what you're saying, but I have to admit,
that sounds so, I don't know . . . kind of
sappy, in a way, when you talk like that."

"Tell me more about that;
what makes it sappy?"

"Something about the attitude, I think."

"And what's the difference
between this attitude
versus Modern Mind's attitude?"

The Seeker thought as he watched bees
above in the wisteria. "It goes back to the
attitude of sacredness versus separation,
doesn't it? Yeah, if I'm separate from this
thing, then your way of speaking sounds
sentimental."

"That's it. But that's exactly
the attitude that gets us into trouble.
Planetary Mind recognizes
the sacredness of the Web of Life
and approaches its lessons
with respect, reverence,
and humility;
having *this* attitude informing
our learning and actions
will help us create more
appropriate technologies."

"How exactly does that work?"

"We'll get to that in a moment.
    Let's first look at some
    characteristics of technology.
    Give me some examples
    as to what technology
    enables us to do."

                        "We don't have to do as much physical labor,
                        because our machines do it for us," said
                        the Seeker. "We can grow more food, travel
                        more easily, build better houses . . . I can
                        frame and roof a house much faster with a
                        nail gun than I can with a hammer."

"Yes, in many ways,
    technology makes life easier.

"It also makes us more powerful
    on the battlefield
    and enables some
    to more easily subjugate
    large numbers of
    less advantaged people, yes?"

                        "Well, yes, it can be used to hurt others, too."

"With technology we can
    create amazing works of art
        and awful weapons of war;
    grow abundant food
        and lay prairies to waste;
    create ample housing
        and raze entire forests;
    uncover deep mysteries
        of the Infinite
        and destroy ourselves
        with astonishing swiftness.

"In essence, technology
    is the great magnifier,

magnifying our individual
    and collective power,
magnifying both
    the great creativity
    and the great destructiveness
inherent within human beings.

"Technology is also the great separator.
    What might I mean by that?"

                "Separator . . ." replied the Seeker, "meaning
                that somehow technology separates us from
                each other."

"Tell me more."

                "Technology separates us," mused the Seeker.
                "How about this . . . going back to cars: they
                separate us from each other. We can move
                around more easily, so we do—and in the
                process we're separated from our families,
                separated from our neighbors . . . most of
                the suburbs I've helped build feel pretty
                lifeless, because you have to get into a car all
                the time. Nobody goes outside."

"But even so, couldn't you make
    the argument that technology can also
    bring us together—
        we can call our friends and family,
        even send electronic
    messages to them?"

                "Yeah, but you also get people only talking
                to others who are just like themselves. The
                Internet's segregated people as much as it's
                brought them together."

"A good point,"
    nodded the Sage.

"Okay, let's grant that technology
          can indeed separate us
          from one another."

                              "What were you thinking of?" asked the
                              Seeker.

"I was thinking about
          all of this instead,"
she said, waving her arm at
the patio they were sitting on.
"Look at this patio, this table
          and these chairs. Look at
          this building, and think about
          all the others in this city,
                    and all the roads.
Where did all the material
come from to build
this city—the gravel, oil,
cement, and wood?"

                              "From mines, forests, and oilfields."

"But which ones? Which forests?
          Which mines and oilfields?"

                              "Timber comes from all over—a lot from the
                              Pacific Northwest, a lot from overseas. The
                              oil's pumped from here in the states, Africa,
                              the Middle East . . . I couldn't tell you
                              exactly where." He thought for a moment.
                              "Same thing with the gravel and cement.
                              You'd have a hell of a time trying to trace
                              where exactly any of it comes from."

"Precisely. Because of the astonishing
          technology used to mine, cut,
          process, and transport these materials,
                    you *can't* really know for sure.
          In terms of technology

being the great separator, then,
what I was thinking was this:
        technology separates us from
                the immediacy of our connection
                with the natural world.
This would not be the case if
you came from the villages of
my great grandmother in Africa
        or my great grandfather in Thailand.
You would know exactly
where the wood and thatching
came from to build your shelter,
        using what Modern Mind
        calls 'primitive' technology,
        however ingenious
        that technology might be.
Modern Mind's technology
        separates us
        from the Web of Life."

                        "Well, okay, but, to be honest, do any of us
                        really want to live in a village in a Third
                        World country?"

"Depends on the village . . .
        and, in any case,
I'm not advocating
        we give up all technology.
        I certainly am not going to.
My point is that Modern Mind's
        blind acceptance of technology
                comes at a price.

"When I have a car
        and well-maintained roads
                and drive to
        an air-conditioned office or school,
I don't have to pay attention

to the weather.
When I wear high-tech shoes,
    I don't have to pay attention
    to the path I'm walking on.
But when I don't *have* to pay attention,
    I *don't* pay attention,
and I lose something precious:
    one more strand of my connection
    with the Web of Life,
    and with the Infinite.

"This disconnection is more than
    a romantic notion.
We turn on the faucet and do not see
    the depleted groundwater
    and the species and cultures harmed
        when rivers are dammed.
We fill the gas tank and do not see
    the drilling rigs, pits, and oil spills
        damaging forests, plains, and oceans,
    nor the habitats harmed and people displaced
        by unscrupulous operators.
We buy our electronics and do not see
    those of us who are practically enslaved
        to mine rare-earth metals
            used in these gadgets,
    nor those exploited in their manufacture,
    nor the enormous piles of outdated gadgets
        that are often 'recycled'
        by being shipped overseas to places
    with lax environmental laws and oversight.
We sip our coffee and do not see
    the plantations destroying songbird habitats
    nor the workers paid pennies for their labor.
We buy food in the market and do not see
    the tragedy of soils and prairies destroyed,
    of animals raised in horrifying factories,

of spiritless, poisoned food,
of empty barns,
    empty towns,
    and empty hearts
    as industrial farming
    displaces family farms."

        "Technology separates us from our actions,
        then?" asked the Seeker.

"From the *consequences* of our actions,
    from recognizing what we're doing.
When we fly, drive, or use any kind
of fossil fuel, we don't *see*
the consequences of our actions,
    even though we, too, suffer
    from these consequences—
albeit much less so
than most people in the Third World.

"So technology can separate us
    from any feeling
    of responsibility for our actions,
can separate us from the Web of Life,
    and, as you point out,
can separate us from each other.
And this separation,
    coupled with technology's vast
    magnification of our powers,
makes for a dangerous combination."

        The Seeker sat pondering. A lavender
        blossom drifted down from the wisteria as
        the bees worked above.

"Tell me what you're thinking,"
    asked the Sage
after a few moments.

"I guess I agree, for the most part . . . I mean,
yeah, like I said, the Internet can segregate
people, but now I'm thinking of what
happened in Egypt and other places, where
technology helped bring people together,
helped them get rid of dictators."

She pondered. "Hmm.
    Yes, it's certainly both,
    but . . ." She fell silent again.

"Perhaps what's muddying our thinking
    is that science and technology
    can do wonderful things for us.
    They extend our physical senses
    and perceptions of the Universe;
    we can no more *feel*
        carbon dioxide levels rise
        or sense the global loss of habitat
            than we can feel
        Earth speeding through space.
    Science and technology have helped us
        to extend our senses in a way,
        to create ways of sensing
        the dangerous signs all around us,
    just as we have used them
        to figure out that Earth is round—
        even though our physical senses alone
        could never perceive that she is round—
    and have used them to lift ourselves off Earth
        and to turn around in space
        to see just how beautiful she is.

"Let's restate it this way, then:
    technology can separate us
    from the consequences of our actions,
    from the Web of Life, and from each other,
    but it isn't always a separator,

nor the *only* separator.
The inequities in our societies—
    the way things are set up
    that dictate who gets to share
    in the bounty of the Web of Life
        and who doesn't—
    are as much separators
    as any technology can be.
We therefore need to be alert
    for this tendency,
    be mindful of the consequences,
    and choose wisely."

"So technology isn't good *or* bad. Technology
is neutral, right?" said the Seeker.

"Hold on; I still don't think
    it's that simple
The statement 'technology is neutral'
    is another story, almost a mantra
    recited by Modern Mind.
But technology is *not* neutral;
every technology has *inherent*
social and political consequences
we would consider good or bad."

"Wait a minute. It feels like you're going back
and forth again."

"Let me give an example.
Say we want to generate electricity.
Think about using nuclear power
    versus solar panels, for example.
If a society were to choose
    to use nuclear power,
    what would it then
    have to do to implement
    such a technology?"

"It would have to build power plants. It
would need to train people to build and to
run them."

"It would in fact need to train
     a highly educated,
     even elite, cadre of scientists
     and technologists to create
     and run these plants.
     What more? What about safety?"

          "You'd have to build a lot of safety
          precautions into it. Even then, they could
          still blow."

"Yes, you would have to train
     yet another elite cadre of safety
     professionals to try to prevent
     these catastrophes and
     deal with them
          as they inevitably arise.
          You would also need
          a strong security force
          to protect the plants.
     And the lessons from the Fukushima disaster
          in Japan all too clearly show
     how the governments and corporations
          overseeing the handling of any disasters
          are deeply tempted
     to minimize or distort the actual danger—
     and how quickly a sense of trust
          in them can evaporate.

"All of these facts point to the absolute
     requirement for centralized
     command and control,
     with all of the political
     and social consequences
          of that choice."

"And solar panels would not have these same
consequences," said the Seeker.

"No, they wouldn't, although you cannot
    ignore other problems
        inherent in solar panels.
    But they do not have as much need
        for scientific elites and security forces,
    because solar panels,
        while still requiring
        technological knowledge
        to create and manufacture,
            do not have the same needs
            for maintenance and protection.
    You could make the argument
        that they are inherently more
        decentralized and democratic.

"Embedded within these technologies,
    and any technology itself,
    are consequences that
    are anything but neutral."

"So solar panels are better than nuclear
power."

"Be careful . . . better to make
    that judgment after examining
    the positive and negative consequences
        of any technology.
    And though I might agree with
    your statement, what is more
    important is *how* we reach any
    such conclusion."

"I think I'm getting it—it's both, then; it's
the inherent stuff about technology *and* also
the way we've set things up that gets us into
trouble."

"Yes. Let's look at another example,
        in this case, creating a new drug.
        How is that decision made?"

                    "Umm . . . well, by the demands of the
                    marketplace?" ventured the Seeker.

"Be more concrete than that.
        Which people might be involved
        in the decision to create
        a new medicine?"

                    "Business people, research scientists,
                    inventors, doctors."

"No one else?"

                    "Perhaps some government workers . . . that's
                    about it, isn't it?"

"It is. Do you see
        any problem with this?"

                    The Seeker thought for a moment. "No, not
                    offhand."

"Isn't it strange that
        the deliberation over new
        technologies is seldom carried out
            in civil debate?
        Or even more rarely carried on with
            a dialogue and deliberation
            that includes all parties
            the technology might affect?

"Instead, because of the way
        we've set things up,
        such decisions are made
        behind the doors of board rooms
        within universities and corporations,
        often under financial pressures

that are difficult to ignore;
   these people cannot carry on unless
   they have reasonable certainty
   that a new drug will make a profit.
And they will also be tempted
   to emphasize only the gifts
   technology has to offer,
      ignoring, and even suppressing,
      its limitations and dangers.
Such motivations, while they
   are amazing and powerful forces
   behind creating innovative things,
cannot be the *only* motivations."

                    "I see what you mean," said the Seeker. "It's
                    beginning to feel like things are set up to
                    bring out the worst of technology."

"Not a bad way to put it,
   and there's an important nuance
      in your statement:
   we're not talking about
      reacting against *all* technology.
What Planetary Mind calls for
   is changing the way things are set up
      as we create technology,
   creating more mindful ways
      in which to use it
   and bring it into being."

                    "So we have to change how we think about
                    creating technology," said the Seeker. He
                    tapped the side of his coffee cup with his
                    fingers and then picked it up and took a sip.

"That's part of the challenge
   of the Great Transformation:
   to develop a sense of sacredness
   that *informs* our use of technology,

which then helps to change
    the way we've set things up
so that *everyone* affected by any technology
can ask important questions
as we consider using it, such as:

    What are the underlying stories
        that drive the desire
            to use this technology?
        How harmonious are these stories
        with the Web of Life?

    What will be the social, political,
        and cultural consequences
        of using this technology?

    Who does the technology
        serve best? How equitably
        will it be shared?

    Who shall represent the interests
        of the whole of humanity
        and the Web of Life,
        as we deliberate
        the use of this technology?

    If, after it is brought into the world,
        this technology proves itself
        to be harmful, can it be corralled easily,
        or recalled and abandoned?

"Presented well, questions like these
    will help Planetary Mind
        to harness the wisdom
        of the collective
    and to consider more fully
    the consequences of technology,
    and help us create and use
        more appropriate technology."

The Seeker thought this over as he gazed
at water gently dripping from the potted
plants. "What about the question, 'What
looks more like a picture of your own soul?'
Wouldn't that help, too?"

The Sage smiled and nodded.
   "Excellent idea.
   That'd be a great question
      to add to the dialogue."

She drew in a breath and let it out.

"And we'll know we're well on our way
   toward integrating an attitude
      of reverent technology,"
         she added,
   "when a team of scientists pauses
      to offer a moment of gratitude
      before beginning an experiment,
   or when operators on a factory floor
      stop to offer thanks
         to the Web of Life,
            to Earth and all creation,
      for the bounty
      that has been given to them."

The Seeker frowned. "Well, there's that
sentimental attitude again. When I picture
that in my mind, I can't help but shake my
head. It seems pretty unlikely."

"Yet the opposite attitude—
   that *none* of these is worthy
   of our respect and reverence—
has driven us into deep crises, yes?"

"Well, yeah—but even so, I bet a lot of
scientists would frown on the idea of
praying to some kind of God."

"I'm not suggesting that.
    One doesn't have to believe
    in any sort of supernatural essence
        to offer gratitude
        to this amazing, mysterious
        process of creation,
    a process that evokes awe, wonder,
    even gratitude, if you only stop
        long enough to contemplate it."

                The Seeker nodded slowly. "I see what you
                mean . . . but that's a pretty big shift."

"It is; let's start, then, with
    imagining just such
        expressions of gratitude
    happening in these unlikely places
    and work to help bring them into being."

THE FORMIDABLE ALLY

A bee landed next to the Sage's cup. She
watched as it investigated the cup and the
saucer, then flew back to the blossoms above.

"This leads to a related story
   told by Modern Mind,"
      said the Sage.
"Imagine this: how might you feel
if you were suddenly stranded
deep in the middle of a forest?"

          "Not at all happy," said the Seeker. "Cold,
          hungry, alone . . . definitely in big trouble."

"Even though you likely would be
    surrounded by food
    and everything you needed
    to make shelter,
if you only knew what to look for.
But even if you had this knowledge
    and were able
        to take care of yourself,
what do you imagine life
would be like in the wild?"

          "It'd still be a struggle, constantly looking for
          food, storing it away from raccoons or bears
          or whatever else might be out there."

"Exactly. That is the sense
    that Modern Mind holds
    toward the Web of Life:
        one of mistrust, at best,
        and hostility, at worst.
        It sees a Web of Life
        that is actively
           and decidedly unfriendly.

Why else would you want to
    'conquer' nature?
You don't conquer something friendly.
Modern Mind clings to civilization,
    believing that it would perish
    from direct exposure
    to the Web of Life.

"The story Modern Mind tells itself
    is that life in the wild,
  in the heart of the Web of Life,
    is alien and hostile.

"Certainly all creatures must expend
    effort within the Web of Life:
songbirds, ants, coyotes, mice—
    all must continually be alert
    to find enough food and avoid being eaten.
But Modern Mind only tells a story
    of a terrifying moment-to-moment struggle
      to eke out a meaningless existence.
This story scares Modern Mind into believing
      that it is much better
      to be severed from the Web of Life,
      to control and conquer it,
    to avoid the capricious
    and inimical nature
    of the Web of Life."

                "Yeah, that's it," said the Seeker. "If you're in
                the wild, you'd better watch out, or the wild
                will kill you."

The Sage laughed.
    "Oooo, yes, indeed it will!
    But is this always true?
Is the Web of Life always unfriendly?
Indeed, is the Universe unfriendly?"

"Well, the Web of Life provides food, shelter,
and all that, and you've said that the Infinite
wants life . . . but if I make a mistake, I'm
dead. I'd say it's both," said the Seeker.

"Exactly. The Universe simply *is*.
The Web of Life
owes life to none
and freely offers it to all.

"If you pay attention,
or 'watch out,' as you say,
if you know what to look for
and stay connected with the Web of Life,
it will more than provide for your needs.

"I would even say that
the struggle to commune
and work *with* the Web of Life,
rather than being something to avoid,
is *essential* to a fulfilling life,
essential to creating beauty and joy;
struggle, as is thriving, is *inherent*
in the greater mystery of the Infinite.

"On the other hand, if you *don't* pay attention,
don't know what to look for,
or are locked into a certain way
of seeing things,
you'll perceive the Web of Life
as a poor provider
and something to struggle *against*.

"As we've seen, much of our technology,
while it serves us,
also dulls our attentiveness;
it replaces our old knowledge
of how to thrive within
the Web of Life,

which then causes us
    to be anxious indeed
    if we're ever stranded
in the middle of an unknown forest,
even in the midst of its bounty."

                    "So we've been dumbed down," said the
                    Seeker. "That's why it feels unsafe to be
                    stranded in a forest."

"Well, 'dumbed down' sounds
    a bit harsh, doesn't it?
Penguins lost their ability to fly
because their environments changed,
    either due to migration
    or to changes in climate,
and they adapted to the changes.
    I wouldn't say they were
        dumbed down.

"Technology, in a sense,
    changes our environment,
    so we adapt, and remarkably well . . .
    but yes, our ignorance
        of the lessons of the Web of Life
    puts us in a precarious position
    when our technologies fail
    or do harm to that which we ultimately
        depend upon for life.

"But back to the point: Planetary Mind
    sees the Web of Life
    neither as enemy *nor* friend,
        but as a process
        that demands deep attention
        and intimate familiarity
        to thrive within it.

"When you are familiar
    with the intricacies of your niche
    and pay attention to it,
        the Web of Life,
            instead of being a powerful enemy,
        becomes a formidable ally.
As we deepen our connection
    through this attentiveness,
    it becomes more than an ally;
        it becomes clear that
    the Web of Life is integral,
    like the air we breathe.
We are it, and it is us."

                The Seeker sat back, taking this in. A
                small puff of wind stirred up the smell of
                evaporating water from the potted plants.

The Sage closed her eyes and breathed in the
scent. They sat quietly in the warmth of the
patio.

"So, then," said the Sage, after a few moments,
"these are a few of the stories
    that, in my view,
    we would do well to rewrite
    as we create Planetary Mind:
        where we learn
to look past our differences
    and harness our common humanity;
    where we see
that the Web of Life is sacred,
worthy of our reverence and respect;
that technology is best harnessed
    with that underlying sensibility;
and that we are meant to thrive
    within the Web of Life;
that it can be our ally and teacher."

"And the overall story changes from 'we are
separate from the Universe' to 'we are a part
of the Universe,'" said the Seeker.

"'We *are* the Universe'
is perhaps a more poetic way to put it.
And changing this story will
do as much to change
the way things are set up
as will changing laws or regulations."

The Sage stretched, sighed,
and smiled.
"Shall we leave it here
for now?"

"I think so. You've given me plenty to think
about."

The Sage laughed.
"Good. I would hate to disappoint."

THRIVING WITHIN THE WEB OF LIFE

Modern Mind sees the Web of Life
    as a heartless enemy
        to struggle against.

Yet look at life.

    Look at the leopards in the wild:
    They are not starving, constantly on the edge of death.

    Look at the crows as they cavort and roost:
    They are not emaciated, with nothing to eat.

    Look at the deer in the fields:
    They do not quake in constant anxiety,
        fearing the hunt, struggling to find food.

    Look even at the insects that crawl:
    They do not hang on to a threadbare existence,
        struggling from one scrawny meal
            to the next.

Certainly,
    all must be alert and aware.
    Certainly, there is hunger,
        even starvation, at times;
    there is great pain,
        suffering, and death.
    But there is also pleasure,
        contentedness, and birth;

        life is all of these.

And the leopards, the crows, the deer, the insects,
    even creatures too small to see
    do not merely survive—
        they *thrive*.

For the Web of Life throws us not
    into a war of all against all,

but calls all
into a creative dance
        with all.

If a species finds life a constant struggle
        and is barely able to survive from one day
            to the next,
            even in the best of times,
        then that species,
        at the first sign of lack,
            shall disappear from the Web of Life.

From its emergence to its extinction,
        each species has its bright moment
            in the Web of Life,
        harnessing the Infinite's drive toward life,
            learning to thrive in the best of times
            so that it might endure
                times of hardship.

So, too, your oldest ancestors,
            deep within the Web of Life,
        did not merely survive;
        they thrived, thrived beyond imagining.
                For how else could humanity
            slowly entwine itself
                within the Web of Life
            in nearly every place on Earth?

The Web of Life is not a force
        challenging us to battle,
    but a sacred and powerful partner
            calling us
            to join in the dance.

PERSONAL
STORY

"Yard work time in about an hour?" asked
the Seeker's wife, poking her head through
the bedroom door.

Yard work time was something they did
together once a week.
They took turns deciding what to work on,
and today it was his turn.

> "Okay," said the Seeker. He was lying in bed,
> reading a book entitled *Unraveling Whiteness*.

She watched him as he read. "How's that
book? The one she suggested, right?"

> "Yeah. It's a real eye-opener."

She thought for a moment, then said, "You
know, I wonder . . . I guess it's still a big
problem, but it's gotten better, right?" She
paused. "I mean, like back in high school—
I just don't remember any racism against the
Asian kids or the African-American kids."

> The Seeker put the book on his chest, and
> frowned. Then, after a moment, he looked
> at her, and raised his eyebrows. "Well, yeah.
> To be honest, when I was in high school, I
> sure didn't see any sexism going on."

Her head gave a little jerk backward. She
looked at him with a frown, then bit her lip,
looked away, and said, "Oh."

        The Seeker gave a small nod. "Yeah."

◆

About an hour later, the Seeker's wife was
sitting in the chair in their bedroom, tying
on her old shoes as they got ready for yard
work time.

"What *also* strikes me is how some of the
phrases you've told me she uses sound, well,
kind of weird," she said.

She stood up and started down the hall.

        The Seeker followed her, buttoning up his
        favorite work shirt. "Which phrases?" he
        asked, as they walked toward the front door
        and went outside.

"Well, 'Web of Life,' for instance. And
'bounty' is another one. I get what she's
trying to say. And, you know, I agree with
it. But can you imagine saying stuff like that
out loud to other people?"

        "How funny." He raised the garage door.
        "We talked about that last time we met, how
        it sounded a little sappy."

"That's it, something like that." She watched
him as he rummaged through the gardening
tools.

"It's like how you're not supposed to show
your feelings in public. Saying 'bounty' or
'Web of Life' sort of shows you care. It has

that same awkward feeling to it. A lot of
people would ridicule you for saying things
like that."

"Right. It'd be strange, hearing some scientist
saying, 'We need to protect the bounty of
the Web of Life.' Something like, 'We have
to secure nature's resources,' sounds more
normal." He picked out two weeding tools.

"But 'nature's resources' sounds so sanitized,
like it's had something stripped from it. It
doesn't have any heart," she said.

She took the weeding tool he handed her.
"So what're we working on?"

"I'd like to weed out the clover. The lawn's
starting to look like hell again," said the
Seeker, walking out to the front yard.

She said nothing, following him. As he knelt
and began to dig out a small clover plant,
she sat down cross-legged nearby. But she
didn't start to weed. She sat, thinking.

The Seeker noticed her. "What?" he said.

"I'm just thinking about what she told you,
about how we need to pay attention to
the Web of Life." She shook her head.
"See? See how funny that sounds?"

He nodded. "Yeah, it does." He dug the
weeding tool underneath a clump of clover.

"Anyway, it reminds me of what I learned last
Wednesday in gardening class."

"You mean the 'woo-woo' gardening class?"
he teased, as he pulled out the clover.

"Speaking of ridicule!" She picked up a small
pebble and threw it at him playfully. "Be
nice!"

    "Hey, you're the one hurling rocks." He dug
    out another plant, adding it to a growing
    pile of weeds. "So what did you learn?"

"The teacher talked about how we tend to
fight nature instead of work with it. He used
lawns as an example, how we try to force
things, mowing and weeding and spraying
to try to make everything perfect.

"I mean, look at what we're doing, pulling
weeds again, like we do all the time, like we
think we can fight them and win."

    The Seeker stopped and sat back on his
    heels. "What are we supposed to do?"

"You're supposed to pay attention. You pay
attention to what wants to happen, and
work with that, rather than try to copy some
picture out of a magazine that has nothing
to do with the Sun, the rain, and the climate
that you're in.

"It's like when we go hiking up to the
waterfall. The forest looks nice all by itself.
Nobody has to go in and weed it all the
time."

    "So what does he suggest about the lawn?"

"Get rid of it," she said.

    "Get rid of the lawn?" he asked, frowning.

"Cover it over with mulch, let the soil build
up, plant some trees, some edible plants,
set it up so it looks more like a little forest

instead of a monocrop. Then instead of
fighting what wants to happen, you work
with it. I think it's a good idea."

The Seeker tapped the weeding tool on the
ground, thinking. "I don't know . . . I kind
of like the green space. And it's a place to
play, too."

"But you're not really losing green space, just
changing the way it gets green. And nobody
plays on the front lawn. Ursula's got plenty
of space to play with her friends in the back
yard. I'd like to try it. It'd certainly be more
interesting than doing the same old weeding
all the time."

She thought for a moment.

"Like that question you learned: What looks
more like your own soul, right? We pull
and cut and fight . . . wouldn't you rather
spend all that time helping something grow
instead?"

The Seeker chuckled and nodded. "Yeah, I
guess I would. You've got a point."

◆

"Where did you come up
        with that response,
    what you said to your wife
    when she made the comment
    about high school and racism?"
        asked the Sage.

"I honestly don't know—it just sort of came
to me. I mean, given what little I've learned,
I was pretty confident that the racism was

there in her high school, but rather than
argue that point, this thought just popped
into my mind, and I said it."

"Interesting. Perhaps from
     an intuitive source, yes?
          I like it.
     And you're right, it's difficult to argue
          with someone who isn't ready
          to hear, but a story is different.
     Your story very gently but powerfully
          brings home the point."

They were seated again near the children's
playground in the park. A handful of kids
were screeching and laughing as they played
a game of tag around a play structure
painted in bright colors.

"And I also like that analogy
     your wife makes about
     helping something grow
     instead of fighting it,"
          said the Sage.

"The way we landscape our homes
     does indeed reflect how our battle
     with the Web of Life is unquestioned,
          even unconscious."

                    "She was curious about what you'd think,"
                    said the Seeker. "I'll tell her you agreed with
                    her. She'll like that," said the Seeker.

The Sage smiled.
     "Will the two of you
     make this change
     to your front yard, then?"

"Yeah. It's less wasteful and makes less work
for us to boot. And she's really getting into
it, so why not?"

"Yes, why not? Going along
   with her energy for the project
   is actually another example
   of working with
      what wants to happen,
      rather than fighting it."

"Okay . . . although I guess I'm a bit . . . well,
wondering what the neighbors will think.
They're going to think we're pretty strange,
even though this project makes sense."

"And that bothers you."

"A little bit. Won't stop me from doing it, but
yeah, a little bit."

"How does that feel in your body,
   the image of neighbors
   pointing and whispering?
   Close your eyes and feel this."

The Seeker closed his eyes, drew in a breath,
and let it out.

"A little queasy in my gut, like I'm about
to be laughed at. Like when I was a kid.
It's going against the grain, against what
everyone else is doing. We'll stand out.
That's the part that bothers me."

"Good, very observant."
   The Sage rose from the bench.
   "Come with me,"
      she said.
   "There's something
   I'd like to show you."

THE RIVER OF MIND

The Sage led the Seeker along a dirt path to
the creek near the playground. They came to
another bench on the bank of the creek.

"Notice the water as it flows,"
        said the Sage, as they sat.
    "I like to think my mind
            is similar to this stream,
        with the water
                shaping the creek bed,
        and the creek bed
                shaping the water."

                            "And the water is the mind?" asked the
                            Seeker.

"Not quite. 'Mind' isn't really a 'thing.'
    Mind is less of an object
            and more of a process.
    Remember when we talked about
            how the creative force of the Infinite
            is like the *shape* of the water?
    The mind is like that, too.
        The creek bed and the water
        together create
    the waves, whirlpools, and eddies
        of the stream
        as the water flows."

                        "So what is the water?"

"The water is more akin to
    the energy that flows through
    your body."

                        "And the body is the creek bed?"

"Yes, and each
        shapes the other;
    energy and body come together
        to create the river of mind.

"Your stories, then, about what
    your neighbors may think
    about you and your wife
    getting rid of your lawn
    influence your body,
    and these bodily feelings loop back
    and reinforce your worry.

"The thing is, you're assuming
    your neighbors' reactions will be negative,
    which may be another example
        of your habitual thinking.
    But remember how your
        coworker echoed
        your sentiments about
        the construction waste?"

                        "Well, yeah."

"You might be pleasantly
    surprised by your neighbors, too.
    Give them a chance."

                        "Okay."

"Let's look more closely
    at another situation.
    Close your eyes and think
    of some recent event
    at your work."

                    The Seeker closed his eyes. After a moment,
                    he said, "Got it."

"Describe to me what you're
    thinking."

"It's about one of my coworkers getting upset at a mistake someone made. I wasn't directly involved, but it was pretty unpleasant."

"Ah, good! Really picture this.
    Try to see sights, hear sounds,
    everything you were aware of
    surrounding this event."

"Okay." He sat silently.

"Tell me now,
    as you're thinking
    about this event,
    what emotions are
    you experiencing,
    how does your body feel?"

The Seeker frowned. "Not happy . . . a little tense." He moved his legs a bit. "Yeah, it feels tense."

"Now try this:
    Replace this thought and recall
        a positive event at work.
    Focus on this
    for a few moments."

The Seeker sighed deeply. He drew several breaths, then smiled.

"What is happening now?"
    asked the Sage.

"I'm remembering a time when we were celebrating after we finished up a project—a building for a community group. Some of us were gathered around a table having a beer, talking about it." He smiled again. "That was a good time."

"Tell me more
about how this feels,
how it feels in your body."

"Kind of pleasant. It's almost as if I'm there."
He moved his legs again. "I don't feel as
tense."

"Good, very good.
Open your eyes."

The Sage nodded.
"You said you felt as if
you were there,
and, in a sense, you *were*.
Your brain perceives
imagined sights and sounds
in the same way it perceives
actual sights and sounds,
and thus interacts with your body
as if such imaginings
were actually happening.
This is why your body feels
the way it does in this moment."

"Hmm. Okay, my thoughts affect my body."

"And vice versa. Now
how did these thoughts
get in there
in the first place?"

"Well, you told me to think them."

"Hah, but of all the situations
that have happened recently,
these are the two
that came into *your* focus.
How, then,
did these thoughts get there?"

"I guess I put them there."

"Precisely. You *chose* these thoughts,
    and they affected your well-being,
        even your body.
    Chosen or not,
        the thoughts you have
        affect your energy and body.
    Repeatedly choose or allow thoughts
        that say 'good job!'
    and your body will, in a sense,
    form itself in such a way
        to favor such thoughts.
    Choose or allow
        angry and roiling thoughts,
    and it will favor those.
        And so on."

"So it's like getting into a habit."

"Well, to a large degree, but then again,
    sometimes there are deeper channels,
        deeper defense mechanisms,
        that we create merely to survive,
        and we really don't know how
            things could be different;
    sometimes it is these mechanisms
        rather than mere habit
    that drive the quality of your thoughts."

"And they're harder to change than habits?"

"They can be quite challenging.
    So be gentle with yourself
    if the course of your thoughts
        falls into old patterns.
    It takes time for rivers
        to change their courses."

WATCHING AND CHOOSING STORY

The Sun grew warm. The Sage watched as a
leaf floated along the current of the stream.

> "Did you have a tough time changing your
> thoughts?" asked the Seeker.

"I was pretty lucky;
> my family was a good
>   and loving place
> to cultivate habits of thought
>   that were helpful to me,
> and was there to help me through
>   the traumas I've encountered.
> And perhaps I was simply
> blessed to have a proclivity to dwell
> on the positive side of things."

> "Interesting word, 'dwell.' Like where you
> live."

"Ah, yes, very clever—and very true,"
> said the Sage, nodding.
> "In any case, it served me quite well
>   as I entered into a society
>   where I looked and talked
>   differently from what was
>   considered normal.
> Even so, it still takes practice;
>   I still have to work at it.

"Now, in your case,
> perhaps it was upbringing,
> perhaps it's a proclivity of *yours*,
> but as I've said before,
> you seem quite comfortable
>   with your darker moods—
> which points to another variable

in my metaphor of
the water and the creek bed:
        the environment in which
        the stream flows.

"What would happen
        if someone dumped poisons
                into the water?
How might that change things?"

                        The Seeker thought. "The water might be
                        bad to drink, but it wouldn't change the
                        flow, really, would it?"

"If it killed off the plants holding
        the bank together?"

                        "Oh. Well, yeah, that would change things.
                        The banks would erode faster, and that
                        would change the shape of the water." He
                        thought for a moment. "So it's more than
                        just the bank and the flow that make the
                        shape of the water; everything surrounding
                        it also affects it. Kind of like when you
                        talked about the Essence of Relation, about
                        how you can't paint a picture of where I end
                        and begin."

"Yes, a good analogy.
        The difference is, *we* usually
        have a choice in creating
        our environments and stories.

"For instance, do you recall
        your reaction when
        I referred to you as a hero?"

                        The Seeker laughed. "Yeah, I think I said I
                        wasn't really interested in the training." He

paused. "But you had a different definition
of hero."

"I did. But if you held on to
    Modern Mind's definition
    of hero, you would expect
    a hero to be larger than life.
    You would expect him or her
        to go in and change things.
    You would expect him or her
        to save us.
    You might even expect
        to not have to lift a finger
        while he or she saved us."

"What does this have to do with the stream?"

"Modern Mind's notions of heroism
    can act as a poison in
    your own river of mind.
    When your hero inevitably fails,
        for all heroes are human,
        your story will likely cause you
        to lament the sad state of affairs,
            to become disappointed,
            disillusioned, or even depressed.
    But if you choose a different
        story of heroism, one in which
        many ordinary people engage
        together to change things,
            however large or small their acts,
    you would be less likely
        to fall into the darkness."

"So my attitude can be like the pollution in
the water, right?"

"Exactly. Let's look at an
    everyday example—how you at times

'make yourself unhappy.'
In our first conversation,
    you lamented that your work
    felt empty and meaningless.
Since then, as you've expressed
    how you want work
    that is more fulfilling,
    expressed concern for what
    Modern Mind is doing
    to the Web of Life,
your mood is indeed often dark, even
    depressed, yes? Why is that?"

"Because what you've told me is pretty
frightening, and I don't know what I can do
about it, and, yes, I want work that's more
meaningful."

"But why the darkness?"

"Well . . . I'm getting older, and I'm
still wrestling with what to do with my
life, especially with this whole Great
Transformation thing. I want to figure it out,
but I've got a family to take care of. I can't
just quit to go off and try to find some kind
of new work."

"You have no savings?"

"Only a little, and we can't dip into that.
We're already trying to catch up with saving
for retirement."

"Perfect. We could go on, but
    this is plenty of material."

The Seeker frowned. "What do you mean?"

"You're accepting all of your
    statements as if they were true,

now and forever.
Look at them:

Older people should have their lives figured out.
My family will suffer if I quit my job.
If I touch my savings, I won't be able to retire.

We could come up with many more
    of these assumptions, these stories,
        if we kept talking.

"More often than not our stories
    are based not on facts,
    but on *beliefs*, usually *distorted* beliefs,
    and projections about the future.
    Your dark stories reflect both,
    and I'll wager that
        these beliefs and projections
    are like grooves in your psyche
        deeply cut from long ago,
        shaped by events that
            have long since passed.

"So, instead of looking at this time
        as a great adventure,
    a chance to finally arrange your life
    so that you're doing work
    you really care about,
        you habitually return to
        these stories of woe,
    as if you're condemned to wander
        in the valley of darkness!"

                "Wait . . . it's not that simple!"

"Really? Take the first one.
    How many people do you know
    who have their lives figured out,
        and are living blissfully
    without stress? The majority?"

"Well, no . . ."

"So why do you hold yourself
to such a ridiculously high standard?"

"But . . ." the Seeker sighed. He looked
down and then away. He felt a twinge in his
stomach.

The Sage softened.
"Look, I know what you are saying.
You're worried about taking care
of your family, yet you also
want meaningful work that
doesn't destroy the Web of Life,
and you can't see your way through
this dilemma, yes?"

"Yeah, that's it," said the Seeker, rubbing his
stomach and staring at the ground. "That's
it exactly."

"I'm not judging these feelings
as much as I *am* challenging
the beliefs and projections
underneath your stories,
which then *inform* your feelings.
I'm challenging you to look at
what you base these stories on,
see whether or not
they are *facts*
or instead are *beliefs*,
and whether they serve
or hinder you
as you find your path.
I'm also calling attention
to how your past influences you,
whispering to you about
how you don't measure up,

how bad the world is,
for *this* reason and *that* reason."

The Seeker was quiet for a time. "Suppose
I *did* start my own little business, doing
whatever it is I figure out I should do. It
might take too much time before any money
comes in. And I might fail."

"Or you might succeed!
And money might come in
more quickly than expected.
And for that matter, you might
never 'figure your life out,'
and your family might
flourish if you quit your job
and chose a different dance.
You *can't* know for certain
what the future holds for you.
Certainly you don't want
to be Pollyannaish about such
an undertaking, but you *can*
choose your stories
and your attitude."

"That *sounds* easy . . . but, I don't know . . . I
don't think it's that simple," said the Seeker
defensively.

The Sage pondered.
"I can't entirely disagree.
Just because you change
your larger story doesn't mean
all your bad habits
and psychic wounds
will suddenly disappear.
Perhaps a new story
will help heal them more quickly,
perhaps not.

Either way, you will
    still have to tend to them.

"It is not that easy to change
        the grooves in your psyche,
    just as it's not easy to change
        the creek bed.
    But if these grooves are negative,
        then you must learn to deal with them,
            to steer clear of these grooves
            and not indulge them,
    if you wish to feel more at home
        in the world.

"*This,* my friend, is your challenge."

                    The Seeker nodded but said nothing.

They both watched the play of the stream
as it flowed by. Sunlight flickered off the
rippling water.

                    The Seeker picked up a small rock and
                    threw it into the stream. He sighed.

"Tell me what you're thinking,"
    said the Sage.

                    "I'm just thinking about what you said . . .
                    trying to take it in . . . thinking about how,
                    I don't know, how I've wasted *all* these years
                    on *all* these negative stories I've been telling
                    myself."

"And this idea of wasted time,
    this is . . . ?"

                    The Seeker looked at her with a frown, then
                    looked back to the water. He suddenly
                    laughed, and shook his head. "It's another
                    one of my stories, isn't it?"

"Yes! Bravo!"
        said the Sage, smiling
    and applauding.

                        The Seeker also smiled, in spite of himself.

The Sage spoke excitedly.
    "You can, if you wish, brood
    upon how you've been brooding
    for too long—or you might also decide
    to try and make a game of it all.
        You can play with watching
                your thoughts and your stories
        like you watch clouds go by,
        as you lie outside
        on your back on a warm day
                without any agenda,
            without spinning off
    into what your stories mean."

                    "So I can choose all of my stories?"

"Well, there's an important difference
        between stories about the past
            and the future,
        which I've alluded to before.

"Stories about the future—
        what you're going to do,
    or what your wife or daughter
        or your neighbors or anyone
            is going to do—
        take you out of the *here* and *now*,
    whether they're positive, uplifting stories
        or stories filled with worry.

"Nobody really knows for certain
    what will happen today
        or ten years from now.
    You're better off, then, simply being present

with your surroundings
and who you're with,
being present with what needs doing now,
taking your plans lightly,
and letting whatever happens unfold.

"Indulging in stories about the future,
whether they're 'good' or 'bad,'
tends to take you
out of the present,
which doesn't serve."

"But doesn't that fly in the face of a lot of
what we've been talking about, about the
huge crises we're facing? Aren't *those* stories?
Didn't you make predictions then?" asked
the Seeker.

"There's a subtle but profound difference
between getting absorbed in stories
versus paying attention
to what is going on with ourselves
and our surroundings
and responding to what's in front of us.
Yes, I get wrapped up in stories
about the future—we all do—
but part of my practice
is to recognize my emotional reactions
and use them as a cue
to step back into the present."

"For example?"

The Sage thought for a moment.

"For instance, when I hear the science
about the catastrophic amount of plastic
inundating our oceans,
how it kills animals that become entangled
in it or ingest it,

and how it may wreak havoc
    with the food chain,
      it tears at my heart.
I might then start telling myself
something like 'we've got to stop this,
    or we'll destroy the Web of Life,'
a prediction that I simply
    cannot foresee with certainty.

"This is different from being in the present,
    listening to my heart,
    seeing that the way we handle plastic
      is simply wrong,
and then supporting those
who work to bring this
    to public attention
or work to ban throwaway plastic,
perhaps even joining with them,
    if that's how I choose to respond.
Do you see the difference?"

                        "Not really. Wouldn't you still be doing the
                            same things either way?"

"Not necessarily. Indulging in
    my little story about
    what might happen
could very well cloud my thinking
    and cause me to take actions
    that are inappropriate.
And even if I *was* doing the same things,
    I'll likely be much more effective
      if I'm not wrapped up
      in these kinds of stories,
because it takes me out of the present moment—
    and for all I know
    the Web of Life will deal

with this dysfunction in ways
that I could never have foreseen.

> "I think I get it—it's about recognizing and
> responding to a situation without getting
> wrapped up in a big drama, a big story."

"Yes. Whether stories about the future
    turn out to be true or not,
    focusing on them wastes my energy.

"Now, stories about your past are different,
    because they're about events
        that *have* happened.
    You've gone to school, learned to drive,
        built houses, married, raised a child,
        lost and found jobs,
        struggled with your career,
            enjoyed successes,
        and suffered through accidents
            and heartbreaks.
    You carry what you've done
        and what has happened to you
        in your psyche and your body.

"Your stories about what you have done
    and what has happened to you
    can either tear you down
        or uplift you."

> "But doesn't a story about the past also take
> you out of the here and now?"

"It can, if that's all you focus on.
    However, stories about the future
    are about events that haven't happened;
    stories about the past *are* about events
        that have happened,
    that are part of your reality.

"Just as Modern Mind's sense
of separation and ennui
is related to its story
of a dead and meaningless universe,
and Planetary Mind's sense
of reconnection, mindfulness,
and humility is related to its story
of a mysteriously creative Universe,
your attitude in the present is in part
related to your larger story
of who you are and why you're here,
as well as your stories about your past.

"For instance, take your idea
about how you have focused
on negative stories for much of your life,
and how you said this is wasted time.
Indulging in this story would indeed
contribute to a dark and brooding attitude.
But you could also tell yourself
about how deeply you've plumbed
the dark side of your psyche,
how well you've learned its landscape,
how you've moved through it
and come out the other side
a much wiser man.
Your attitude, I'll wager,
would then be much different,
much more even-tempered,
and would serve you well."

The Seeker chuckled. "I like that a lot
better."

"And it's your story to tell, if you wish.
You have more power
over your stories of the past
than you think.

Part of your challenge
is to watch your stories mindfully,
letting go of those focused
        on the future,
and choosing stories about your past
        that serve you.

"Now, I'm not advocating
        that you stuff darker thoughts
        down into your psyche
        without any examination,
                where they might fester
                and become even more powerful.
        If a thought occurs that has new and useful
                information in it, then it's worth examining,
                whether or not it's a dark thought;
                        but if you find yourself ruminating,
                        going over old territory without any benefit,
                        that is something to recognize
                and gently take yourself out of.

"You asked how I can be so happy—
        to a large degree,
        I *choose* it.
        I *choose* to try to keep clear
                the flowing waters of my mind,
                to watch and pay attention
                        to the thoughts I carry
                        and to the songs I listen to and sing,
                to try not to let my stories, good or bad,
                        carry me out of the present moment.
        I *choose* to try to walk lightheartedly
                even with the pain
                of my past and my ancestors' past,
                in a culture that, consciously or not,
                often holds the color of my skin against me.
        I *choose* to try to cultivate uplifting stories
                about what I've done

and where I've been.
I *choose* to try to cultivate
a sense of gratitude."

GRATITUDE

"You know, that's one thing that's frustrating sometimes," said the Seeker. "Why is it that I'm so ungrateful? I mean, really—dirty dishes aren't that big a deal; all the hassles I have to deal with are *nothing* when I think of how many species are going extinct, or how many people on the planet would *kill* to have a house like mine, to have a life like mine," said the Seeker.

"Tell me about these hassles."

"Okay, yesterday after work I had to stop by the doctor's to pick up some paperwork before they closed, and as I pulled in to the parking lot, my wife calls, all upset about something at work. So I have to talk to her, to try to soothe her and help her deal with it, and by the time we're done, the doctor's closed, so that's a wasted trip. And then I have to rush off to pick up my daughter from music lessons . . . on and on and on."

"Drowning in
    the millions of details."

"Uh-huh."

"Like being pecked to death
    by ducks," said the Sage.

"What?"

The Sage laughed.
    "A little phrase I once heard.
    I think the image is hilarious—
    and useful to me."

The Seeker furrowed his brow but had a
small smile on his face. "Really? How's that?"

"I still get caught up in details—
 not as much as before—
 but when I do,
   the image reminds me of my attitude.
 Really, does everything fall apart
 when I don't get around to
 finishing the shopping, the laundry,
   the cleaning?

"There's also another thing:
 what's the common element
 in the details of this story
 you just told to me?"

The Seeker thought for a moment.
"Common element . . . the doctor, my
wife . . . my daughter's practice . . ." He
frowned. "I don't really see one."

"It's *you*.
 You're part of each little detail.
 And what were you telling yourself?"

The Seeker thought for a moment. "That I
have too much to do, that I shouldn't try to
do so much?" he ventured.

"Well, that's probably true, too,
 but what *I* heard was this:
   I *have* to.
 I *have* to go to the doctor's.
 I *have* to talk to my wife.
 I *have* to pick up my daughter.
   This, too, is a habit
   of seeing life as drudgery."

"Well, these things *do* have to get done, don't
they?"

"Yes, but you can try out
    a different attitude of seeing life
        as a privilege
    by telling a different story:
        I *get* to.
    I *get* to go to the doctor's—
        I'm fortunate to even *have* one.
    I *get* to have a wife—
        to talk with her through her troubles.
    I *get* to have a beautiful daughter—
        to pick her up from a practice
            that enriches her.
    I *get* to feel the frustration of living
        in the depths of Modern Mind—
        and find my way through life
        in the midst of it.
    I *get* to feel the pain and heartache
        of the great injustices and
        extinctions we now face,
        and to learn to love the world anyway.
    I *get* to struggle through these realities—
        and figure out what my work
        in the world is.
    I'm *blessed* with this life
        that I get to live out—
        especially with so much bounty
            surrounding me.

"All of this helps lift this small story up
    from one of dark drudgery
    toward one of lighthearted gratitude.

"You may not be able to control
    all your circumstances—
    your genetic inheritance,

your mother and father's parenting,
the way a society has set things up,
        or your doctor's office hours—
but you certainly have control
over your expressions of gratitude."

                    "I'll be damned," said the Seeker. "'I get to',"
                    he said, trying out the words. "Seems easy,
                    but I don't know . . ."

"Oh, yes," said the Sage.
    "You've had a lot of practice
        doing and being otherwise;
        you have indeed
            dwelled in this place
            for a long time,
        have cut some rather deep grooves
            in your psyche,
        which, as we've seen,
        also affects your body.
If you're not careful, you can literally
            make yourself sick
if you indulge these dark thoughts."

                    The Seeker chuckled, shaking his head.
                    "Huh. Yeah, I can see how that could
                    happen."

The Sage nodded.
    "It's not about eliminating turbulent times,
    it's about meeting them mindfully,
        so you can develop a sturdy sense of resilience
        to help you ride through such times
            a bit more gracefully.
    I'm also not suggesting that
    you deny the difficulties that you face
            and stuff down
                legitimate feelings.

It's terribly important to be discerning—
        to bring in expressions
                of gratitude
                as you move through daily life,
        but not use them to mask
                the real feelings that might emerge
                as you square up to your struggles.
And it's important to go easy on yourself
        when you find yourself
        slipping back into old patterns.
Even *that* is an opportunity
        for you to express gratitude
                for recognizing these patterns
        as you begin to master your stories.

"It might not be all that easy,
        but it's well worth the effort,
        for it is in the space of mastery over story
                that people throw off addictions,
                live in constant joy,
        bring down empires,
                get locked up
        or burned at the stake
        or crucified
        or shot, and have
impacts that last for millennia."

                        "Uh . . . that last bit sounds an awful lot like
                        that hero training I didn't sign up for," said
                        the Seeker.

The Sage laughed, slapping her knees.
        "Ah, yes, but nobody said
        it couldn't be fun, did they?"

                        "No, I guess not."

"Might be a good thing to remember,"
        said the Sage, smiling.

FOUR WAYS OF WORKING IN THE WORLD

The Sage rose.
　　"Let's go back to the benches
　　by the playground. I so enjoy
　　watching the children play."

As they walked, the Sage continued.

"Speaking of 'hero training,'
　　that leads to another challenge.
　　One of the fundamental flaws
　　as Modern Mind acts out its story is this:
　　　　as it creates wealth,
　　it damages the Web of Life.

"This creates a dilemma difficult
　　　　for any individual to reconcile:
　　Are my actions, is my work,
　　　　supporting the juggernaut
　　　　of Modern Mind,
　　or helping humanity face
　　　　the challenges of
　　　　the Great Transformation?

"There are only a few instances
　　where one's work creates
　　wealth in a way that not only
　　does no harm to the Web of Life,
　　　　but holds the Web of Life
　　　　as sacred, even as one
　　　　performs the work."

　　　　　　　　"What work does that?" asked the Seeker.

The voices of children grew louder as
they came around the path toward the
playground. They could see the children
running around the play structure,

immersed in a raucous game of tag. The
Sage smiled at their play.

"There is a farmer I know
    who tends a few acres of land.
    As she respects and pays attention
    to the land, the plants, and the animals,
    holding them as worthy
        of her reverence,
  she learns the way forward
in her craft:
      where to plant,
      where to allow the animals
          to graze,
      what the soil needs,
      what the creatures in the soil
          need,
      what the plants need.
The land, the plants, the animals,
    even the worms and microbes
    in the soil
are her teachers.
    And as she practices, she finds the land
        becoming richer,
        and the bounty growing.
Her example demonstrates
    one of the ways
    open to our creativity."

                "Well, that's *nice,* but . . ." The Seeker caught
                himself. "I'm sorry . . . I don't mean to be
                sarcastic—really, that's good for her, but this
                seems like an exception to the rule."

"For the most part I have to agree with you.
    The economy of Modern Mind
        offers few instances where
        one might honor the Web of Life,

even in a small way, through paid work,
    although it's certainly vital
    that we create ways that do so.
For the time being, you might need
    to distinguish your *job*,
      or what you do
      to earn money,
    from your *work*—
      at least the work that
      I'm talking about,
the work of creating Planetary Mind—
      which is where your
      passions intersect with
      what the world needs.
You may need to make peace
    with the reality that
    the economy of Modern Mind
    may have no idea
    how to value your work;
the way things are set up
    might not reward such work."

"My 'work' might be unpaid?"

"It's possible."

"That doesn't help."

They arrived at the bench next to the
playground and sat. The Sage nodded
toward one of the men watching over the
children as they played.

"Let's say that man
    is being paid to take care
      of these children.
    Pretty important work, yes?"

"Yeah," said the Seeker. "But he's not being
paid much to do it, right?"

"Exactly. In this society,
    a lot of important work
    reaps very little financial reward
        and often isn't recognized.
    Indeed, it often seems
        that the so-called 'unseen hand'
        is more attached to
            someone who's oblivious,
        or to sinister or selfish forces,
    than it is to one with concerns
        for the greater good.
    This is what we're up against."

                The Seeker pondered. "It sometimes seems
                pointless to try."

"Ah, but you told me,
    in so many words,
    that your heart aches
    to do work that speaks
    to your spirit, yes?"

                "Oh, yeah—no question about that!"

"Remember how I said that
    if I want to change things
    I'd best allow my heart to
        break open?
    It's not enough to leave it there
        once that happens;
    then all you have is a broken heart.
    It's vital to find
        that *engagement* with the world
    that takes you out of your
    petty interests and worries
    into something much larger
        than yourself.

"And just because Modern Mind
    might not value such work,
    it doesn't mean you have
    to take its story to heart."

                    "But here we go again, talking about work,
                    and I still don't feel any closer to having
                    an answer. I know that no one's going to
                    give it to me, but I'm still feeling like I'm
                    floundering here."

"Maybe that's because you're
    still not ready to let go of the idea
    that it must be paid work?"

                    "Well, I have a mortgage and bills to pay; I
                    need medical coverage for my family . . ."
                    He grimaced, and rubbed his stomach. He
                    dropped his hand and sighed.

The Sage looked at the Seeker
    and cocked an eyebrow.
"Yes, and in *your* case,
    changing your circumstances
        might be rather hard;
    your job isn't all *that* bad,
        and it pays well.
You could stay with this job indefinitely.
Even though your spirit calls to you,
    and you feel a certain urgency
    about the world and your work in it,
it still might take something
    to shock you out
    of your comfort zone."

                    The Seeker bit his lip. "I can't argue too
                    much with that."

"Well, don't condemn
    yourself either.

Perhaps all you truly need
      is a bit more time.
And I'm not saying that
      all worthwhile work is unpaid.
But as we change the way things are set up,
and collectively learn to support work
that truly serves humanity
      and the Web of Life,
we still have to deal with the reality
      of the way things are set up *now*,
      and figure out how to make our way
as the old economic system crumbles,
      and a new one arises."

                    "Right, so what do I do about that?"

"You get to create a combination
      of the two that works for you
            and your family.
You might decide to reduce your
            expenses so that you can
            reduce your hours at work,
                  if that option's available.
You might find out that there's work
      available at your job,
                  such as finding ways to recycle
                  the construction debris
                        at your job sites.
You might break away and
      create your own business
      that focuses on green
      projects and construction.
It all depends upon
      what calls to you."

                  "Look, you've said that a few times, 'what
                  calls to me.' Right now, I'm just not hearing

anything. Can we focus on how you figure
out your calling?"

"Okay," said the Sage.

She got up, walked over to a tree not far
from the playground, picked up a stick,
returned, and sat back on the bench.

"Here's one way you can think about it."

She leaned forward, drew a square in the
dirt in front of the bench, and then drew
horizontal and vertical lines through the
square, making four small squares.

"Mind you, I always hesitate
        to use such models,
        because Modern Mind loves
        and sometimes worships
            its models and structures.
        Life, just like the Web of Life,
        is much more organic
            and chaotic than that.
        But that doesn't mean
            these models aren't useful,
            it just means that
            it's okay if things don't always
                fit our models.
        Use them as you would any tool
        and then put them back on the shelf
            when you're done.
        Don't worship the tool."

The Seeker laughed. "Okay, I won't," he said.

"What this model looks at
        is *how* you do things,
            not *what* you do,

but looking at your work in this way
can help clarify what that work is.

"Now, the upper two boxes,
      on the right and left,
            represent the person.
      The bottom two
            represent the group, or society.

"The left boxes represent the interior world,
      the right the exterior world.
      The upper left box, then, represents
            a person's interior world.
            What might I mean by that?"

                              "Person's interior world—that's sort of their
                              mental state, their thoughts. And emotions,
                              too, right?"

"Yes, as well as the person's stories,
      and all the mysteriousness
of the subconscious and the unconscious,
      which some consider the soul and spirit."
      Now, what about
            the upper right box?"

                              "The person's exterior world—that's more
                              what they have, their stuff, the people in
                              their world, their job, that sort of thing."

"Close—I'm thinking more
      of their *behavior* in the world,
      as the person relates to all
      these things you list.
      This represents how he or she
      acts in the world.

"Now, we could launch
      into a whole exploration
      of what happens when

the interior world,
the person's story,
collides with his or her behaviors
in the world, causing all sorts
of drama,
but perhaps another time."

She pointed the stick
at the lower right box.
"What does this one represent?"

"That's the group's exterior world." The
Seeker frowned. "I'm not sure what that
means, but if it's like the upper one, does it
also have to do with behavior?"

The Sage smiled. "Excellent!
I hadn't thought of it that way.
This box represents
the systems of the group—
that is, the way things are set up:
laws, rules, regulations, norms,
all of which, now that you mention it,
are very much indeed the
group's behaviors."

She pointed at the lower left box.
"And this?"

"The group's interior." The Seeker frowned
again. "Can the group *have* an interior? That
doesn't make sense."

"Ah, but it can—this represents
the *culture* of the group:
its norms, ideas, beliefs, archetypes,
its collective consciousness and spirit—
and its overarching Stories."

She sat back on the bench, and twisted the
stick in her hand.

"Again, we could have a fun
        conversation about what happens
        when *these* two collide,
        about the craziness and dysfunction
        that happens when
                the stories and the systems
                don't match up with each other,
                or when they come up against the realities
                        of the lessons of the Web of Life.
        And Modern Mind, at this point,
                is particularly dysfunctional,
        but as we've seen, dysfunction is part
                of the dance of the Infinite."

                                "Okay," said the Seeker. "So if we're not
                                going to talk about that, what's this model
                                for?"

"I use it as one way
        to answer your question:
        how do you discover your calling?
        Not so much *what* your work is,
                but *how* you express your work,
                        your gift to the world.
        The four boxes point roughly to four ways
                of working in the world, and I use
                them to frame four questions:

Is your calling more in alignment
        with the 'person's interior' box,
                akin to a philosopher, nun, or monk,
        exploring the depths of one's
        psyche and connection with the Infinite
                and how that consciousness
                        reflects and echoes
                throughout the whole of humanity?

Is your calling more in alignment
   with the 'person's exterior' box,
      akin to a helper or healer—
      learning and teaching about
   lessons of the Web of Life,
      whether through healing work
      or building cooperative structures
      through work in small, local
      businesses or projects?

Is your calling more in alignment
   with the 'group's exterior' box,
      akin to an activist,
   working to change our systems
      and institutions
   to make the *right* thing to do
      the *easy* thing to do—
   the 'right' thing being that
   which is in harmony with
   the lessons of the Web of Life?

Or is your calling more in alignment
   with the 'group's interior' box,
      akin to a storyteller—
   the artists, historians, and scientists
      creating ways of telling
      the story of Planetary Mind?"

                    "Hmm." The Seeker rubbed his chin,
                    thinking. "Yeah, that makes sense. That
                    helps."

"Well, remember, as with any model,
      it *is* a simplification.
But this model raises another question:
      of these four ways,
      which one is the most critical
      as we navigate through
      the Great Transformation?"

The Seeker thought. "I'm kind of wrestling with what you mean by 'critical.' You mean critical for me, or critical for the world?"

"For both, actually."

The Seeker pondered. "It depends, doesn't it? It's different for each person."

The Sage set the stick down on the ground
beside the bench.

"That's my feeling, too,
    but you will run into people
    who think otherwise.

"I know, for example, people who
    feel the most important work
    is cultivating the inner world
        of the person;
    others say it's working to create
        small ventures or projects
        or to help heal others.
    Still others insist it's changing
            the story of Modern Mind,
        and others feel it's changing
            the systems of Modern Mind.
    So who's right?"

"Well, none of them, right?"

"Or perhaps *all* of them, in a way.
    No one really knows,
    and the question is ultimately distracting.
    As we just discussed,
        it doesn't serve to latch tightly
        to *any* story about the future.
    We can spend our time arguing
        over which way is best,
        or we can simply quiet ourselves,

be in the present,
    listen for what our next task is—
then get to work,
    engage our passions,
      offer our gifts,
serve the work that the world needs,
allow for the conflicts we'll feel
    between the inner and outer worlds,
take time to analyze what works,
and dance with the whole
      wonderful chaotic dance.

"And it's absolutely vital
    that we also *support* one another:
that the monk
    support the activist,
that the storyteller
    support the helper and healer,
    and so forth.
Who knows which one's work
    is more important?
Who knows where
    our thoughts and actions
    will echo out to?

"Just like the Web of Life,
    I'll wager that there's no hierarchy
    among these ways,
and that from the work of all
    will emerge profound
    and wonderful new ways
of being in the cosmos."

        "Huh." The Seeker leaned forward. He
        watched two of the children as they chased
        one another around the slide.

The Sage sat quietly.

"Huh," he said again after a few moments, nodding his head. "Yeah, that really helps. I mean, I've always wondered about that saying, 'Be the change you want to see in the world.' It seemed like there was more to it than that."

"Ah, yes: the quote attributed
        to Gandhi.
    Being the change
    we want to see in the world
        is vital,
    but as Gandhi also did,
    we must also *make* the change
    we want to see in the world—
        even though we might
        not live to see this change
        come to fruition."

"And we have to work out of all four boxes to do that."

"Collectively, yes. But everyone
    doesn't need to work through all four ways.
    I happen to gravitate toward
        being a storyteller, which I love,
        though I'm also enjoying learning more
            about helping others
        through my time with you."

The Seeker laughed. "So I'm kind of an experiment."

"Ah, yes—and you are a
    wonderful experiment indeed."

They laughed together and then grew quiet again.

The Seeker shifted on the bench. "Yeah, this
really helps. But I'd still like a little more. I
mean, I'm definitely no storyteller, but how
do you figure out which way is yours?"

"Well, you are a *bit* of a storyteller;
    each of us has some of each
    way within us.

"But as to your question:
    Some people find it useful
    to ponder the question,
    and write down their thoughts
        in a journal.
    Others find help in books
        that guide the inquiry.
    Still others have it seemingly
        thrust upon them
        by circumstances,
        serendipity, or through
        sudden flashes of intuition.
    Some find their way by simply
        talking with and learning from others—
            though as I've said before,
            be careful of anyone
            who adamantly claims authority—
        but you *can* engage in inquiries
        and dialogues that do a delightful job
            of tapping into a collective,
            emergent authority that is different
            from a single authority,
        and can wonderfully enhance
        your individual searching.

"In fact, here's something you could try:
    Remember how I talked
    about finding where your passions
        intersect with what the

world really needs?
You can take a sheet of paper,
and on one half, write down
    what you're passionate about,
    what you love to do.
On the other, write down what
    you see needs doing in the world,
    especially those things
        that break your heart.

"Then, gather five people together,
            and without you saying anything,
                just simply taking notes,
            have them look at what you've written,
            and see what patterns they see.
            They just might come up with something
                you'd never think of.
You could then all talk
    about the four questions,
    and what roles resonate
    with each person.
You could all also ponder the question,
    'What work, what calling, is worth
    devoting one's life to, and why?'
There's a lot of wisdom
    and collective authority
    that can emerge from such meetings
        that might never emerge
    from one's solitary musings."

                    "Kind of like that other meeting you talked
                    about, the one where the people talked
                    about building the house."

"Very much so.
    I would expect many
    useful answers to come out
    of such an inquiry.

"But whatever you end up doing,
   don't worry about doing it right,
   don't worry about doing it wrong.
Just simply work
    with what comes up."

        "Okay, I'm liking this. I'm thinking I might
        get somewhere with this."

"One last thing:
   it might be helpful
   to consider how your work
   may serve others,
      no matter what way
      becomes clear is yours.
How often have you heard
    that helping others
    contributes more
    to our well-being
    than gathering wealth?"

        "I've heard that, but, if that's true, then why
        don't more people do it, if it's more likely to
        make them happy?"

"Do you doubt that
   serving others is fulfilling?"

        "Not really, but why don't we do it more
        often?"

"Venture a guess," said the Sage.

        The Seeker thought for a few moments.
        "Well, we're certainly not encouraged much.
        I mean, we pay lip service to serving the
        greater good, but it's sure not rewarded."

"That's part of it;
   our culture's stories
   glorify our proclivities

of self-interest and greed.
We have both inside of us:
    the desire to serve
and the desire to be selfish."

               "And that's where the idea of survival of the
               fittest comes in."

"Yes and no. Again that phrase
    has been twisted and bent
    toward justifying excessive
    competition in our culture.

"But think back on our ancestors.
    Who might have been more fit:
    the one who looked out entirely
        for his own selfish desires,
    or the one who looked out
        for the clan, as well as
        some of his own desires?"

               "Seems obvious to me: the one who looked
               out for the clan," said the Seeker.

"It probably varied widely
    from place to place,
        and culture to culture,
    but it was likely that those
        who survived and thrived
        were the ones who knew
        when and how to cooperate
        and were loyal to the group.
They then passed along
        an innate tendency toward
        serving both the whole
        as well as themselves.

"In a sense, the Infinite
    put a desire to serve within us,
    alongside our selfish needs

to survive and procreate.
I'll wager that this is why
those who serve others
as well as themselves
and are willing to sacrifice
themselves for others,
are happier and more fulfilled,
more fit, than those who
remain entirely selfish."

"I guess it makes sense. Otherwise, why
would people do jobs in non-profits that
don't pay so well, especially when they have
a choice?"

"Exactly."

The playground began to empty of children
as the day grew long and dinnertime
approached.

"There are some things
I'd like to show you
next time we meet,"
said the Sage, rising.
"For now, it would be
interesting to see what comes up
as you ponder your work.
Play with watching
and choosing your stories."

"Okay, I'll practice."

THE GIFT OF SACRIFICE

A teacher once said of the struggles of life,
   "All this is suffering."

Yet the Web of Life requires suffering,
   for life feeds on life.

The lion must suffer hunger
   when he does not catch the gazelle;
   the gazelle lives by the grace
      of the lion's suffering.

The gazelle must suffer pain and death
   when caught by the lion;
   the lion lives by the grace
      of the gazelle's suffering.

But what is the deeper meaning of suffering?
      Suffering is pain turned inward,
      self-reflective, self-absorbed.
   Yet suffering turned outward is transformed
      into the gift of sacrifice.

Even the Sun—he slowly dies.
   Each second, his flesh,
      four million tons, vanishes
      in blazing fire and light.

If he thought only for himself,
   he would suffer, saying, "I'm dying—
      every second, I grow older;
      my flesh withers away."

Yet, the Sun gives his fire and light to all his children.
   His third child, Earth,
      needs his light to bring life
         to the Web of Life.

Without his dying, Earth could not live.

Thus the Sun's suffering is transformed
    into a cosmic sacrificial gift
that this child might be warmed,
and the Web of Life might thrive.

CONNECTING WITH THE INFINITE

The Seeker stood in the kitchen in his
family's house, staring at another sink full of
dishes.

> He sighed. "Why can't she just do the damn
> dishes like she says she will?" he asked
> himself.
>
> "Why is this so annoying to me?" he said
> aloud. "Why don't we just get a damn
> dishwasher? Although I suppose we'd fight
> about *that*, too, wouldn't we."

The house stood silent. A droplet of water
slowly formed from the faucet, and plopped
into a pot half-filled with gray water.

> He closed his eyes and let out a breath. "I'm
> grateful that I have a woman in my life
> that I love." He opened his eyes and looked
> around, feeling foolish.
>
> He turned back to the sink. "I'm grateful
> that she's willing to give me so many
> chances to practice gratitude." He snickered
> to himself.
>
> He looked at the dishes. "I *get* to wash the
> dishes. I sure didn't predict *that*," he said
> sarcastically. "I'll bet that . . ."

He stopped himself. He stood quietly, as another droplet of water formed and plopped. "My God," he murmured, shaking his head.

He closed his eyes again, and took in a breath and let it out. He opened his eyes and looked at the pile of dishes in the sink. He picked up the pot, upended it into the sink, and set it down. He opened the cupboard below the sink, pulled out a bottle, squirted a bit of soap from it into the pot, filled it with water, and stirred up the suds with a washrag.

He picked up a dish and scrubbed it. He put it to one side of the sink, starting a pile for rinsing. As he washed the rest of the dishes, he thought of a new song he had recently learned, and began humming the tune, trying to change his mood.

"What was it that brought you back into the present?" asked the Sage.

It was late in the afternoon, and they were in a rural park close to the ocean, sitting near a campground in the blue chairs the Sage had brought. Every now and then, the sound of the surf and the smell of salt water drifted through their conversation.

"I'm not quite sure," said the Seeker. He thought back, remembering how he felt in front of the sink. "I think it was mainly because I'd been watching my stories since

our last meeting, trying to see which ones
were the past, which ones were future.
Somehow that worked its way in as I was
getting wound up. I mean, I couldn't believe
how quickly I got upset."

"And what happened
        as you hummed
        the song to yourself?"
asked the Sage.

"I think it helped," said the Seeker, "but I
wonder, can't a song also distract you, take
you out of the present?"

"Good question. I think it can
        all too easily, so it's best
                simply to watch out for that.
        Really, your only task
                is to be present
                to whatever it is
                that needs attention
                        in the moment."

The Seeker raised his brow, shaking his head.
"Yeah . . . easier said than done, though."

"Oh, yes. I constantly
        catch myself, too."

The Seeker leaned forward in his chair, his
hands clasped in front of him. "We also
talked about it, and she apologized—and I
apologized to her, too. I realized how often
I can get really mean. And sarcastic. I *am* so
damn negative, and I don't like myself that
way."

"And you now see that it is
a habitual attitude
fed by a habitual story?"

"Yeah, I do—and it's awfully easy to fall back
into."

"And when I say 'bravo,'
you'll know why, yes?"

The Seeker gave a small huff and smiled.
"Because I recognized it, right?"

"Yes. And bravo!"

The Seeker smiled. "Thanks." He paused for
a moment. "There's something else I did. I
decided to write down a list of things that I
was grateful for. I did it a couple of times."

"What an excellent idea!
Are you going to continue
with this?"

"I think so . . . maybe not every day."

"Quite all right . . . you can change
whatever practices you feel called
to do as you see fit, as long as they
help you pay attention.
You never know which practice
will really click with you,
so it helps to try many."

"One other thing." He stopped, looked
down, and shifted in his seat.

"Yes?"

"It's that . . ." He flushed slightly. "I wanted
to give this to you." He took a little bundle

tied up with multicolored twine from his
jacket pocket and handed it to the Sage.

"My goodness, what's this?"
      said the Sage, her eyes shining.

She unwrapped the bundle and pulled out a
necklace of silver beads and polished stones
of amethyst and malachite.

"Ah, it's beautiful!"
            she exclaimed.

                  "My wife suggested it. She picked it out,
                  actually. Anyway, I'm grateful—we're
                  both grateful—that you've been willing to
                  spend time with me, to talk with me about
                  everything."

The Sage smiled.
      "You're quite welcome.
            I, too, have gratitude for
            your willingness to listen.
      I find each chance to talk
      about these ideas quite precious.
            I'm also learning, remember.
      So my thanks to you, too."

She unfastened the necklace and clipped it
around her neck.

                  "Looks good," said the Seeker.

"Very thoughtful of you both,"
      replied the Sage.
      "Thank you!"

AWE

The Sage bent from her chair to pick up a
dried leaf, withered, crumbling, and faded
to a pale green.

"Look at this leaf,"
          she said.
     "Quite unremarkable, isn't it?
Nothing like this necklace
you just gave me, yes?"

                    The Seeker smiled. "Well, I might have said
                    'yes' before, but, knowing you, I'm pretty
                    sure you have a different answer."

The Sage laughed.
     "Indeed I do."

She handed the leaf to the Seeker.

"Look at it closely.
          Notice the lines, or veins,
     how they branch out from
          the main stem."

                    The Seeker did so, tracing the veins with his
                    eyes.

"Notice its symmetry,
     how each vein occurs
     at regular intervals.
     And as you continue
     to look at this leaf,
     know that it used starlight
          from the Sun
     and stardust forged billions of years ago
          in the form of carbon dioxide
          and water
               to create itself.

It followed a code that evolved
          over eons,
in an unbroken line from
billions of years ago to today."

                    "Interesting."

"More than interesting—fascinating.
     And more than fascinating,
     it is indeed miraculous."

                         "I have to admit, it doesn't strike me quite
                         that deeply," said the Seeker.

"Not surprising. You have been
     thoroughly trained
          to ignore
the constant miracle of the Infinite
that surrounds you,
          and to pay attention instead
to the flashy images, selfish ideas,
and anxious stories perpetrated upon you
          by Modern Mind."

                         "No arguing that," he said. "But it's even
                         more than that . . . it feels a bit childish."

"Of course. And where
     does *that* thought come from?"

                         The Seeker shook his head and chuckled.
                         "Another story from Modern Mind? That it's
                         childish to look around the world in awe,
                         isn't it?"

"Exactly. And that story robs
     you of another habit
     you could practice to build
     your sense of connection
          to the Infinite.
     Perhaps it might be useful

for you to try this:
look around,
   see what miracles surround you,
and say out loud, 'Beautiful.'"

      "That sounds forced, like I'm pretending to
      look at the world in awe. It reminds me of
      telling myself 'good job.'"

"That's quite all right; in fact,
   you can start by simply pretending.
   Intentionally pause for a moment
   and practice exclaiming
   'beautiful' to yourself
      when you see a leaf,
      a small flower,
      your daughter's eyes,
      the graceful motion
         of your pets,
  even weeds growing
      through cracks in the sidewalk.
Take time to notice color,
      shape, texture, movement,
         sound,
         smell,
         even taste—
and pretend, at first,
      to marvel at how each of these
      is part of the unbroken chain
      of life, stretching back
      for millennia.
Notice also hills, clouds, sky . . .
      and reflect that they are all
         made of stardust
         that harkens back to
      the Beginning of All Beginnings."

"Okay," said the Seeker. "It's like paying
attention to stories, right?"

"It is. I find that not only has it had
    the benefit of keeping
        my stories in check,
    but it also helps me to move
    beyond the intellectual *idea*
    of the Web of Life as sacred,
        into a deeper *experience* of the sacred.
    To me, the awe I hold
        toward the Web of Life
        is core to my spirituality.
    And why not?
        Isn't the Web of Life,
        even the entire cosmos,
        where my body came from?"

"I think I see. We're just giving respect to . . .
what . . . the place we came from?"

"*Emerged* from, more precisely,
    just as the pear blossom,
        the fruit, and its seeds emerge
        from the tree that nurtures them,
        which in turn emerges from
        the water, air, soil, and light
        from Earth and the Sun.
    If you try noticing the quiet miracles
        surrounding you and keep up with it,
    I'll wager that you will, sooner or later,
        be delighted when you unthinkingly
        feel the miraculous
            striking you deeply,
        and you find yourself,
    involuntarily and sincerely, blurting
        out 'beautiful!' as you

behold a creation
of the Web of Life."

The Sage rose from her chair, folded it, and
gathered up her backpack.

"Where to?" asked the Seeker, as he followed
suit.

"To the beach. There's something there
I want to show you."

TURNING TOWARD EVENING

They followed a path through the park,
leading to the beach. The Seeker followed
the Sage across the sand dunes and toward
the shoreline. About twenty yards from the
edge of the water, they stopped and set up
their chairs.

The Sage pulled out two blankets and a
wool cap. She handed the cap and a blanket
to the Seeker.

"Put these on,"
        she said,
    "so that you don't catch a chill."

They wrapped themselves in the blankets
and settled into their chairs. The Seeker
pulled on the cap. The Sage motioned
toward the horizon, which had drawn closer
to the Sun as evening approached.

"Watch the Sun
        as the horizon draws nearer to it.
    Take care, of course,
            to not look directly at it
            so you don't injure your eyes.

"Sense the Sun not as a disk in the sky,
    but instead as it really is:
    a fantastically enormous sphere
        of tremendous fire,
    traveling through space,
    ninety-three million miles away.

"Feel yourself sitting on Earth,
    as she rolls gently eastward,
        turning herself in space,

turning away from the Sun,
rolling toward night."

They sat silently for a few moments as Sun
and horizon drew closer together.

"I kind of get this in my head . . . but it isn't
as easy to feel it in my body," said the Seeker
after a while. "It's more of a fleeting feeling."

"Oh, yes," said the Sage.
"It's hard not to see
the Sun as a disk
moving across the sky.
But keep trying."

They sat quietly, as Earth turned
her face away from the Sun.

"I see it now," said the Seeker, as the horizon
began to cover the Sun. "But I sort of have
to work at it." He squinted and relaxed in
his chair.

The horizon hid more and more of the
Sun, until it slowly disappeared behind
the moving Earth, the last bit of orange
dimming and dissolving into the grey
horizon.

"So, then, did the Sun indeed set?"
asked the Sage.

"Well, no; the Sun didn't set . . . Earth
turned."

"Yes. 'Sunrise' and 'sunset'
are outdated terms.
It is just like you said:
the Sun is not rising or setting;
Earth is turning.

She turns toward morning;
    she turns toward evening."

By this time a few stars had begun to appear
in the sky. The Sage motioned toward them.

"We also tend to think of the sky,
    the stars, and the Sun
    as 'up,' yet what happens
    as Earth turns?"

                    "Well, 'up' continually changes, doesn't it?"

"Yes, yet we never get a sense
    of 'down,' do we?"

                  "Not at all."

"Let's try another experiment,"
    said the Sage, getting up
    and unfurling another blanket
    as she walked around the Seeker's chair.
        She spread it on the sand.
"Lie there,"
    she said,
"with your feet pointing to the south,
    your head to the north."

                The Seeker did so.

The Sage returned to her chair and sat.

"Try this. See yourself here,
    on Earth's surface,
    as she continues to turn.
See her turn away from the Sun,
    and see the stars suspended far off.
See if you can sense
    the stars and the Sun not as 'up'
    in the sky, but as 'out' in the sky."

The Seeker was silent for a moment. "Well, I can kind of see this . . . but it's hard to keep it in focus, sort of like when we were watching the Sun."

"That's okay—keep playing with it."

She sat silently. More stars became visible as the sky darkened.

After a while, she said,
"Look up and to your right;
there's the Big Dipper.
Do you see it?"

The handle of the Big Dipper pointed at an angle toward the center of the sky.

"Yes," said the Seeker.

"From your perspective,
tell me where the Sun is now,
and where the Big Dipper is
in relation to your body."

The Seeker thought. "Well, the Sun is over there, below the horizon to the right and behind me, and the Big Dipper is out over to my right and in front of me."

"Wonderful! Notice your words:
'out' and 'over there.'
Normally, if you were standing,
you'd probably say something like
'*up* there.'
Lying down on the ground helps
to see the stars as
'out there' instead of
'up in the sky'
and makes it easier to imagine

the circle of stars around Earth.
Do you sense this?"

"A little bit," said the Seeker. "I mean, I sort
of get the idea, but it's hard to *feel* it."

"Good, yes. Again, this is
an unusual perspective
you are cultivating.
And this is a good reminder
that there are limits to what
we can take in
through these exercises
in our normal waking states—
that we're not *built*
to perceive Earth turning
or to see the stars as 'out there,'
just as we're not built to perceive
the speed at which Earth
travels around the Sun."

"What do you mean, 'normal waking states'?
You mean we *could* perceive these if—what,
we meditated or something?"

The Sage laughed.
"That, to be honest, is something
I'm just learning about myself:
the possibility of other states of reality that
might allow us to break through these limits.
Maybe they can, maybe they can't;
I don't know yet.

"Even so, you can still play with this exercise
throughout your days and evenings.
Notice where the Sun is.
See how much you can feel
Earth turning her face toward
and away from the Sun.

Play with seeing
    where you are on Earth,
    how she turns slowly
    in space around her Sun.
Play with the sense that the Sun is stationary
    in relation to the turning of Earth.
It's a delightful little meditation."

The Sage pulled out a red flashlight and a
star chart.

"I find that being familiar with the stars
    makes it easier to practice this sensation
      of looking *out*, and not *up*,
    into a sky filled with a field of stars.
        Plus it's just fun
      to know the constellations."
And the Sage began teaching him
    the patterns of the stars
      out in the sky.

THE TEMPLE OF THE INFINITE

"Now, those six stars
    form the bow of the archer
        in the constellation
        of Sagittarius.
    Do you see how it
    intersects with the path
    of the Milky Way?"
        asked the Sage.

                  "Uh huh."

"I love this fact . . .
    when you look at those
    particular stars in Sagittarius,
        you're looking toward
        the center of our galaxy."

                  "Wow," whispered the Seeker. After a
                  moment, he added, "Amazing." He fell
                  silent as he watched the sky.

The Sage turned off the flashlight and set
it down, along with the star chart. They
sat quietly for many minutes as the waves
curled back and forth in the dark.

"How do you feel?"
    asked the Sage, suddenly.

                  The question startled the Seeker out of his
                  contemplation. "Feel?"

"Yes. You know . . . emotions—
    sad, glad, mad . . . scared . . . ,"
    said the Sage playfully.

"I know, I know." The Seeker thought for a
moment. "I don't know . . . peaceful, I guess.
Calm."

"What might these exercises
have to do with story?"

"Story? I don't know." He caught himself. "I
bet you're thinking I should try harder."

The Sage chuckled.
"Yes, please."

The Seeker let out a breath, thinking. He
looked out into the sky. "The story of
Modern Mind . . . it says we're separate, that
this is all an accident." He was quiet again
for a few seconds. "But I don't really feel
that . . . this all doesn't *feel* like an accident,
not when I'm out here looking up . . .
looking *out* into the sky."

The Sage smiled in the darkness.
"No, you are not an accident."

◆

"Here is something
I've been pondering lately,"
said the Sage after a few moments.
"Tell me what you think.

"We make meaning, make story,
simply by being alive, yes?"

"Yes," answered the Seeker.

"Each of us
creates our own story,
just as the apple tree
creates blossoms and bears fruit.

The parent raising his children,
    the woman driving a forklift,
    the criminal hardened by
    years of incarceration,
    or the nun who serves in the slums:
    all of us create and desire story.
But where does this instinct,
    this desire, this craving,
        come from?"

        "Maybe it happened when humanity came
        along."

"I wonder. To me, that sounds
    awfully reminiscent
    of Modern Mind's story,
        that humanity
        is the pinnacle of evolution.
Perhaps the desire for story
    is an emergent phenomenon,
but then again, might this instinct
    for story emanate from a deeper source?
Might all beings
    within the Web of Life
    desire story in their own way?
Might we all have emerged from a Universe
that somehow, in a way
    beyond our understanding,
        also craves story?"

        "You said that the Infinite is both the painter
        and the canvas being painted upon," said
        the Seeker, musing.

"From the moment
    the Infinite called the Universe forth
    out of the Unfathomable Mystery,
    It has created, destroyed,
        and created once more,

as if It knew what It was doing.
And through its creations,
   the Infinite seems to strive,
   to crave something,
   perhaps to know Itself
      more clearly,
      more deeply."

          "But there's a part of me that still feels
          skeptical that the Universe is somehow, I
          don't know . . . conscious?" said the Seeker.

"I'm not surprised. Don't we go back
   and forth between the realms
   of intuition and evidence?
   We struggle as
      our intuition awakens,
   bringing forth ideas,
   thoughts, and feelings
   that our logical minds
   have no idea how to handle
      and will reflexively deny."

          The Seeker sat quietly, listening to the waves.

"No, not an accident,"
   she said after a moment.
"In a very real sense,
   you and I *are* the Universe
   looking back into itself,
   contemplating itself after nearly
   fourteen billion years of creation."

          "Huh," said the Seeker.

"Huh, indeed. And perhaps
   this is why you feel peaceful?"

          "Maybe so," said the Seeker.

"I think so," said the Sage.
     "I like this part of the story."

                              "I do too," replied the Seeker.

They sat quietly again.

◆

Two water birds floated past in the dark,
calling quietly to each other as they followed
the shoreline. The breeze shifted, blowing at
their backs and toward the sea. The waves
swept in and retreated several times before
the Sage spoke again.

"When we first met,
          you told stories of unease,
               anxiety, and estrangement.
          Your stories reflected
               the wider story of Modern Mind;
               you sought greater connection
               and meaning within your relationships
                    and in relationship
                    with the Infinite,
          just as Modern Mind, in its own way,
               seeks reconnection
               and meaning within its own
               relationship with the Infinite."

                         "I have to admit, when I think about it, I still
                         get knotted up about things."

                         The Seeker looked out at the stars. "Though
                         I'm better at watching the stories I'm telling.
                         And I don't feel as anxious as I used to."

"Ah, that is good to hear,"
          said the Sage.

She was quiet for a moment.

"In the temples, mosques, and churches
        of Modern Mind,"
            she continued,
    "we go to recite and seek solace
        in words of belief.
But I think we are now entering a time
        when belief alone will no longer quell
            our estrangements and anxieties,
        will no longer satisfy our souls.
I think we are collectively awakening
to a deeper yearning for connection,
        and will seek instead
greater *communion* with each other
        and with the Universe,
        outside of our buildings,
        where we can lie down
            or stand up, play,
                look around,
            and say 'beautiful.'

"Perhaps the Infinite waits quietly,
            waits for us
    to reawaken and find our way
        out of our temples and stories,
    to come back to the home
        that both surrounds us
            and is within us,
        back to the home
        that was there all the time."

                    "And that's what we're doing here?"

"Yes . . . I thought it might be fun
    to bring you here, to show you.
    I find it easier to feel this connection
        in places like this.
    But I believe, with enough practice,
        it can be done almost anywhere.

Perhaps our new temples
  will be everywhere.
Perhaps Planetary Mind will hold
  the entire Universe as holy—
    all space, all time."

                    "The temple is everywhere," pondered the
                    Seeker.

"Yes, for doesn't the creative
  force of the Infinite
    flow everywhere?"

                    "Even in a junkyard? Even when we go to
                    war against each other?"

"Of course. How could it be otherwise?
  Just as we have emerged
  out of the Infinite,
  and we express the Infinite
  through our actions
    and creations,
  so too are these actions
    and creations—
      even those we recognize
      as absolute desecrations—
  a mysterious part
  of the whole, holy story."

                    "It's just harder to fathom in the middle of a
                    slum."

The Sage smiled.
  "Or in the middle of a clear-cut forest
    or a country overrun by warlords.
  But it is still there."

                    The Seeker picked up some sand and ran
                    it through his fingers. "*That's* the hard
                    part, seeing it in those places. I wonder if

it's really possible." He looked again to the
night sky.

"We may dump poisons
    and pollution,
    create rubble- and
        blood-strewn battlefields,
    strip minerals from within Earth,
        and level the forests on her hills,
    but in time, over centuries and millennia,
        the Essences of the Web of Life will
        assert themselves, will seek
        a new balance with what
    we have wrought upon her."

                    "Whether or not we awaken."

"I think we will;
        as I sit here in this temple
        and look out at the decorated sky,
        listening to the chorus of waves
            and shorebirds,
        I remember.

"I remember that the Infinite
        has been driving Itself
        to greater order
            and beautiful complexity,
        to greater consciousness,
            for billions of years.

"I remember I am bathing
            in this cosmic stream
            that carries me
            on this delightful journey.

"I remember that we are much larger
        than we think;
        we are not
        meaningless specks

on an insignificant planet
in a vast, heartless universe;
we are instead dazzling flashes of brilliance
on a tiny but delightful,
sacred sparkle of stardust
called Earth.
The Infinite could not be
one millimeter smaller
and still be able
to call us into being;
the Universe worked patiently,
for billions upon billions of years,
to create us.

"There is a place for you in
the Universe, in the Web of Life.
You never were separate
from the Infinite."

She fell silent once again,
and they sat basking
in the starlight.

◆

After the waves broke and retreated several
times, the Seeker said, "One of these times,
I'd like to hear more about the Essences of
the Web of Life."

"Ah, yes. They're fun
to ponder and play with.
But another time."

Just then, a wave came up, broke, and
surged toward their chairs.

"Ah! Ooo!" cried the Sage,
quickly rising, laughing

as the wave
threatened to inundate them.

The Seeker scrambled up, grabbed the
blanket and his chair, and moved them back
from the water.

"Yes, another time,"
said the Sage, with a loud,
happy sigh as
the wave retreated.
"It's getting late."

They packed up the star chart, the flashlight,
the chairs, and the blankets.

"Getting very late indeed,"
she said,
their feet sifting through the sand
as they walked up toward the dunes
and disappeared
into the darkness.

THE SMALLEST VESSEL

What is the smallest vessel that can hold a human being?

Certainly it is more than the skin and bones that contain
    the pulsing of the individual life within;
    one human cannot forever stand alone and separate.
    Even the wise woman who lives in the forest
        apart from others
      serves as the wise woman for those others.
The smallest vessel that can hold a human being must include
    at least one other human being.

But two humans cannot forever stand alone and separate.
    They need young ones
      to raise and teach,
      to help with the daily chores,
    and, finally, to take charge
      and carry on
        as the elders grow old, their bodies dying,
        releasing their starlight
        and becoming stardust once again.
The smallest vessel that can hold a human being must include
    at least the family.

But the family cannot forever stand alone and separate.
    It needs others to help in the gathering of food,
      the building of shelter,
        and in caring for those who are sick or hurt,
      just as it helps others in their own time of need.
The family needs others to bind together with
      in times of catastrophe,
        of want, and of war,
    as well as to rejoice with
      in times of plenty, and of peace.
It needs others to share in the knowledge of Earth's gifts
    and to learn the ways of the wise old ones.

The smallest vessel that can hold a human being must include
at least the clan.

But the clan cannot long survive alone.
It needs oxygen to breathe, food to eat,
and waters to quench its thirst.
It needs medicines to heal those who are sick.
It needs insects to pollinate and clean
the forests, savannas, deserts, and prairies.
It needs jaguars, hawks, turtles, sparrows,
trees, flowers, vines,
and all manner of animals and plants
both seen and unseen
to teach the wordless songs of the Infinite.
The smallest vessel that can hold a human being must include
at least the whole of the Web of Life.

But the Web of Life cannot long survive alone.
It needs a Mother,
willing to share her flesh:
air,
water,
the makings of soil,
and the mixing together of life-giving elements,
so that the Web of Life might form itself
out of her own body.
The smallest vessel that can hold a human being must include
at least Earth herself.

But Earth cannot long survive alone.
She needs a star to draw light from
to warm her creations,
to cause the clouds to form,
the winds to blow, and the rains to fall.
She needs a Moon
to steady her
as she dances spinning through the seasons
and to cause her oceans to pulse

with life-giving tides.
She needs planets, comets, asteroids,
to pull and push, and sometimes collide with her
and stir the cauldron of creativity.
The smallest vessel that can hold a human being must include
at least the Sun and his children.

But the Sun and his children
cannot have come into being alone.
They need a galaxy of stars,
forming, living, dying, exploding,
creating the elements for life.
They need a billion seeds,
a billion possibilities,
and the death of the Grandmother Star
to bring forth that one precise possibility
that allowed our Sun to be born
and his children to emerge.
The smallest vessel that can hold a human being must include
at least the galaxy.

But the galaxy cannot have come into being alone.
It needs forces, particles, and fire,
spinning forth
from the first callings of the Infinite,
forming into billions of colossal galactic clouds
spiraling out into the primordial cosmos.
The smallest vessel that can hold a human being must include
at least the Universe.

But the Universe cannot have come into being alone.
It needs an Unfathomable Mystery,

a time of no time,

a place of no place,

a Beginning of All Beginnings,

so that the Infinite can then call forth the Universe,
and the Universe can explode into being.

Therefore . . .

The smallest vessel that can hold a human being,
   that can hold you yourself—hold all beings—must include
   the whole of the Infinite . . .

      at the very least.

EPILOGUE

Once, a reporter
    who learned of the Seeker's heroic works
    asked if it was really worth it,
        trying to help heal such a tiny part
        of an enormously damaged planet.

"Oh, yes—yes it is," said the Seeker, smiling.
    "It most certainly is worth it.
        For me, it's the best
            work in the world.
        And who knows where
            it will echo out to?"

The Seeker stepped out of the office, empty-handed,
    for in his shock,
  he had left his notes behind.

He walked to the bus stop
  and numbly boarded the bus home.

Images flashed over and over in his head:
  the meaningless numbers on the report,
  the look on the doctor's face,
  her stunning words of the prognosis—
    months, perhaps a year at most;
  the ridiculous suggestion
      to find a hobby, so as to distract
      from the pain—and the reality;
  the distant efficiency of the staff
    as appointments with the oncologist
    were coolly made.

He got off the bus at his stop, but,
  not ready to face his family,
  walked in the direction
  away from his house.

He wandered down streets and away from houses
  and found himself at the creek
    not far from his home,
  a creek abused with litter and debris:
    an old refrigerator,
    several tires,
    an abandoned shopping cart,
    piles upon piles of litter.

A raven roosted in a leafless tree
  on the opposite shore, watching him.
  Water trickled from somewhere,
  and a thin membrane of oil
  rode on top of the water
  as it oozed slowly past.

But the Seeker neither heard nor saw
    any of this.
    He stared vacantly, thinking
        how his family would fare
        and how he would miss friends,
    and felt a cold terror
        beginning to rise.

The raven called twice and flew away.
    The Seeker watched as it disappeared,
    then turned back to stare sightlessly
    at the creek.

He wondered, disoriented,
    what to do, what to think,
    and was afraid to feel . . .

And through his questioning, terrified heart,
        words
    found their way to him:

    Find a place on Earth, even a small one,
        that is in need of care.
        Take care of it.
        Learn what you can from it.

And his heart changed, ever so slightly.

He broke his stare
    and saw the badly abused creek,
        as if for the first time.
    His gaze fell on the abandoned refrigerator,
        and a thought fell into him.

He scrambled down the bank
        to the refrigerator,
        lying on its side near the water.
    He stood below it, and put a foot
    on its edge, moving it, testing its weight.
    He bent down, ignoring the dirt and mud

that soaked into his shoes
    and the dull pain that murmured within.
He dug his hands under its edge,
    and lifted it up and over.
It landed with a thud,
    slid a few inches back down the bank,
    and stopped.

The Seeker dug his hands in
    and again heaved it over.
It caught the edge of a sleeve,
which tore as it landed again
with another thud.

Over and over he lifted and heaved
    in spite of the pain just below his sternum,
    slowly working the refrigerator
up the bank of the creek,
    doors flopping open and closed,
until he reached the top.

He sat on the refrigerator,
    lungs laboring from the effort.
The unfamiliar fatigue he felt frightened him
        far more than the pain.
    He noticed blood running
        from a cut on his hand.
    A deep welling inside him broke loose,
        and he began to sob.

◆

The following day,
    after he cried much of the night with his wife,
    after he huddled together with her and their daughter
        for most of the morning,
    after he felt the shock of the news reverberate
        through his circle of friends,
        family, and coworkers,

and after a long conversation with the Sage,
　　the Seeker returned to the creek
　　　　in an old white pickup truck.
　　　　A quiet sense of purpose
　　　　　filled his eyes.

He backed the truck up to the
　　refrigerator and opened the tailgate.
　　He put on a pair of gloves, walked
　　behind the refrigerator, lifted one
　　end of it, flipping it over
　　and onto the back of the truck,
　　leaning against the tailgate.

He bent down, lifted,
　　and shoved the refrigerator
　　into the bed of the truck.

He made his way down the bank of the creek.
　　He noticed an old tire half buried in the mud
　　and tugged at it.
　　The tire moved a little, with a slurping
　　sound of water, but did not come free.
　　He took the gloves off
　　and tried to dig around the tire with his hands,
　　but it was sunk deeply into the mud,
　　and he gave up.

He picked up several old cans and plastic bottles,
　　walked back up the bank, and threw them
　　into the back of the truck.
　　He made several trips up and down the bank,
　　until he grew tired.
　　He covered the load with a tarp,
　　and drove off.

◆

The Seeker returned the next day
   with a shovel, rope, and boxes.
   He dug out several tires, and threw them
   into the back of the truck.
   He filled boxes with piles of trash,
   hauling them to the truck.
   He tied one end of the rope to the shopping cart,
   the other end to the truck, and pulled the cart
   up out of the creek.

Over and over, the Seeker went
   up and down the bank of the creek,
   until he grew exhausted.
   He drove off, returning a few hours later,
   and again filled boxes with debris
      until he again grew exhausted.

The creek sighed with relief
   as each load of debris
      was removed.

In the weeks following, the Seeker
   hauled truckload after truckload
   of debris out of the creek.

He brought family, neighbors, and friends
   to help with the tasks.

Between visits to doctors' offices,
   he huddled close to his wife and daughter,
   and he worked at the creek,
      feeling a sense of urgency,
      of time drawing to a close,
         as energy ebbed from his body.

He continued to help the creek
   restore itself as best he could.

In the months following,
    when the Seeker was not with his family,
        he spent his time with the creek.
    He learned much about what
        must be done to help a creek
        bring itself back to life.
    He befriended those whose ancestors
        had lived nearby for millennia,
        learning of the history of the creek,
            of his home.
    One of the descendants he met
        was learned in the plants of the bioregion.
        She gave him advice and helped him
        as he planted seedlings of species
            that once thrived
            on the banks of the creek.
    He brought more volunteers
        to help with plantings.
        As some of the seedlings did well
            and others did not,
            he and the volunteers,
        watching and listening,
            began to know where to plant.
    He gathered people together
        to speak with leaders
        in local and state governments,
        to gain their support,
        to enforce and create rules
        prohibiting the dumping
        of pollutants into the waters.

And he remembered
    that there was larger work that needed doing,
    remembered to support others' efforts
        to change the way things were set up.

He gave time and money to organizations
    seeking to restrict
        the power of those
        responsible for damaging
            the Web of Life
    and to place upon these powers
        greater responsibilities toward humanity
        and toward Earth.
He brought friends
    to workshops put on by
    two of the volunteers,
    two women who worked
        to build bridges
    between the classes, races, and sexes.
He taught children
    who came from nearby schools,
    telling them some of his story,
    showing them what he had learned,
        showing them
        how to help the creek,
    and encouraging each to find his or her role
    in creating the story of Planetary Mind.

He helped his friend next door
    bring neighbors together
        over food and drink,
    for more than just celebrations,
    but also to help each get to know one another,
        share skills and stories,
        and build stronger ties
        within the neighborhood.

Once, a reporter
    who learned of the Seeker's heroic works
    asked if it was really worth it,
        trying to help heal such a tiny part
        of an enormously damaged planet.

"Oh, yes—yes it is," said the Seeker, smiling.
   "It most certainly is worth it.
      For me, it's the best
         work in the world.
      And who knows where
         it will echo out to?"

Time passed, and the creek
   showed signs of renewed life:
   birds, fish, frogs, snakes, mice—
      and even a fox returned.

The Seeker sometimes sat quietly at the creek, alone,
   noticing the flow of mind and water,
      listening and watching,
   and smiling as he caught himself
      sometimes murmuring "beautiful."

As he spent
   more time with the creek,
      learning from the creatures and plants
         that lived in and near the water,
      listening to the birds singing
         the stories of the day—
   more time with the ocean,
      smelling and tasting
      the lessons of the waves—
   more time with the hills,
      watching Earth as she turned evening
      and looking out into a sky filled with stars—
   more time with his family,
      weeding through their fears,
      confronting their upsets,
      healing old wounds
         and facing the abyss together—

his sense of urgency and being afraid
gave way, every now and then,
to feelings of acceptance
and of peace.

A slow, deep shift had begun inside of him,
even as the pain of the disease
would sometimes flare . . .

and then subside.

ACKNOWLEDGMENTS

In order of appearance, more or less.

First and foremost, to my parents and siblings, who had to put up with the odd and sometimes quiet number three son and his wild ideas and strange wanderings;

To the colleagues and teachers of all the circles I have roamed through over the decades, especially RPR and NLD, my early mentors, who believed in me and were also the first people I met who helped me begin to see beyond the story of Modern Mind;

To the people I met through the New Road Map Foundation, especially Vicki Robin, Monica Wood, Peter Mui, and Monique Tilford; to the people of the Foundation for Global Community, especially Richard Rathbun; to the people of Bioneers, especially Nina Simons and Kenny Ausubel; and to the people of The Pachamama Alliance, especially Lynne and Bill Twist— it was they, as well as other people in like-minded organizations, who introduced me to many of the ideas of the leading Cultural Creatives that are now bringing forth what I have termed *Planetary Mind;* and to Bill and Lynne Twist again, for their early support of my efforts in writing this book;

To Duane Elgin, the Godfather of this project, whose encouragement, cajoling, and audacity helped me begin to set these words down on paper;

To those authors and thinkers whose works informed the content of this book—and while I'm grateful for their insights, I must also acknowledge that all errors of interpretation are mine alone, and my building on their foundation in no way means they agree with or endorse this work . . . in fact, I expect some scowls, crossed arms, tapping toes, and demands for explanations;

To Chris Cone, who early on believed in the project and offered vital encouragement, criticism, and suggestions to the first manuscripts;

To those who read early, chaotic drafts, saw the possibility within them, generously spent many hours helping me bring forth the structure of the book, and helped me hold to deadlines: Tina Poles, Richard Senghas, Geof Syphers, and Terry D. Taylor;

To the many people who read later drafts of and offered encouragement and feedback on the manuscript as it evolved: Tom Atlee, Laura Elizabeth Baker, Chris Bell, Virginia Bertelsen, Tree Bressen, Rana Chang, Mark Dubois, Peggy Duvette, Sierra Hart, Jess Johnson, Aaron G. Lehmer, Linda Milks, Shirley Murphy, Maluhia Pacal, Diane Poslosky, Larry Robinson, Kay Sandberg, Laird Schaub, Nick Setka, Veronica Stone, Steve Spanier, Karen Kingsley Swett, Rick Theis, Ruel Walker, Lindsay Whiting, Marcin Whitman, and Robert Zeuner; special acknowledgments go to Sharon Preston, who constantly gave me encouragement over the years;

To Lee Glickstein of Speaking Circles International, speaking coach extraordinaire, who helped me become a more effective public speaker as I presented drafts of the book to groups;

To my wonderful editors, Becca Lawton, Nancy Lewis, Jordan Rosenfeld, Regina Sara Ryan, and Jennifer Tucker—their combined insights, gentle criticisms, suggestions, and sharp eyes made the manuscript so much stronger. And to Jordan Rosenfeld again, serving as my "book coach," who helped me bring the manuscript to completion and talked me down from the tree on more than one occasion;

To the fantastic people at Book Passage in Corte Madera, CA—especially Leslie Berkler and Nick Setka—who are all amazing and helpful professionals for authors and book lovers; also, to the many pros and seminar teachers at Book Passage's classes as well as at the events held by the Book Publicity Forum in San Francisco (formerly NCBPMA) and the Bay Area Independent Publisher's Association, who have been so generous with their publishing knowledge and experience;

To my "Publishing Pros," who shepherded me along the path of producing the book and bringing it to market: Alice Acheson, Vicki DeArmon, Arielle Eckstut, Joel Friedlander, Suzanna Gratz, Lorna Johnson, and Karen Leland;

To Daya Ceglia, David Kerr, and Claudine Mansour, for their delightful and uncanny ability to take my ideas, concepts, and feelings and translate them into beautiful images;

To those who encouraged me to take retreats and opened up their homes to me, where I was able to step away from day-to-day living and make critical progress on the manuscript, including Anita Gamba, Col. Jim Wood, US Army (ret.), and the "Gardening Gals," Christel Casjens and Susan Osofsky;

To those who offered their ideas, support, and encouragement over the years, including Jeff Aitken, Vinit Allen, Tracy Apple, Mark Bachelder, Connie Barlow, Steve Bhaerman, Ellen Bicheler, Mary Carleton, Sunny Chayes, John Crowley, Drew Dellinger, Natalie Doel, Michael Dowd, Ian Dunbar, James Dunbar, Chuck Durrett, Lindsay Dyson, April Eberhardt, Dave Ellis, Amber Espinosa-Jones, Lloyd Ferris, Karen Harwell, Anne Hillman, Emily Hittle, Betty-Ann Kissilove, Eve Libertone, Jon Love, Valerie Love, David Lukoff, Jeni Lyon, Steve Lyman, Peter Mattair, Kathryn McCamant, Deb McMurray, Scott Meinzen, Stacey Meinzen, Alex Morrison, Monique Muhlenkamp, Melissa Nelson, Phyllis Oscar, Shannon Laliberte Parks, Patrick Picard, John Renesch, Neal Rogin, Michael Stone, Jon Symes, Bruce W. Thompson, Mary Reynolds Thompson, Tatiana Tilley, Shannon Tilton, David Tucker, David Usner, and Pat Usner;

To those associated with Dominican University of California's First Year Experience/Big History program, including Mojgan Behmand, Jaime Castner, Cynthia Taylor, and Neal Wolfe, both for encouraging me to offer my work as a tool in the field of Big History, and for allowing me to present portions of the manuscript to Dominican's First Year Experience classes;

Finally, to my teacher, YKF, who has helped me to begin to see the aliveness and enormity of the seen and unseen Universe—and to begin to more gracefully square up to my struggles and cultivate humility, confidence, delight, and peace as I move through life's dance:

Thanks to you all. I will never be able to repay you. I feel deep gratitude for your generous, gracious, and wise spirits.

QUESTIONS FOR FURTHER EXPLORATION

The following questions are taken from the *Reader's Guide to The Holy Universe*, which is available online (along with the *Teacher's Guide to The Holy Universe*) at www.theholyuniverse.com.

You may also, if you wish, visit www.theholyuniverse.com to explore discussion and dialogue groups, organizations, contacts, ideas, and other sources of information and inspiration that you might find useful as you discover your own dance within the Great Transformation.

PROLOGUE

1. To learn about the importance of story, the Seeker, at the Sage's suggestion, throws a birthday party for a friend. How does your experience with celebrations compare with the Seeker's celebration? When have stories been a part of the celebrations you've attended? Do you agree that adding stories helps celebrations? How might you incorporate stories into your celebrations?

2. The Sage's gender is not revealed until the middle of the Prologue, and her ethnicity is not revealed until Part I (and it is only implied at that point). How much were these revelations a surprise to you, and why? How does your response fit (or not fit) with the larger story of your culture? What assumptions did you make about the identities of both the Sage and the Seeker in the absence of such telling details?

THE GOSPEL OF THE UNIVERSE

3. The Sage points out to the Seeker that he's thrown away the stories he was given as a child, but can't reconcile with the "Big Dumb Rock" story—to which the Seeker exclaims, "I don't have a story!" What times in your life, if

ever, did you *not* have a larger, overarching Story? What happened to cause you to lose your story? When, if ever, was your story seriously challenged, even changed? What happened to challenge and change your story?

4. How does the arc of the story of the Universe that the Sage presents—from the Beginning of All Beginnings to the creation of the Web of Life—compare with the story that you grew up with? What parts of the Sage's creation story, if any, resonate with your story?

5. The Sage uses the terms "Infinite" and "the creative force of the Infinite" to describe her idea of ultimate reality. What words do you use to describe this ultimate reality? What do you believe are its characteristics?

THE EMERGENCE OF THE WEB OF LIFE

6. The idea of catastrophe spurring creativity recurs often in the story of the emergence of the Web of Life. How have you used the large and small catastrophes in your life to create anew? How differently do you view these catastrophes now versus when you were in the middle of them?

7. At the end of this chapter, the Seeker answers the Sage's question about the emergence of consciousness with, "When humans came into the picture, that's when we became aware." The Sage, however, counters with a much larger view of consciousness. Do you agree with her view or the Seeker's view? How far do you think consciousness extends throughout the Universe? Throughout the *history* of the Universe?

THE BOOK OF ANCIENT MIND

8. The overarching Story of Ancient Mind tells a story of deep connection with the Web of Life. Are there times in your life in which you felt more connected with the Web of Life? With the Universe? What needs to happen for you to feel that connection? How easy or difficult is it for you to invoke this sense of connection?

9. The Sage points out that many (though not all) cultures of Ancient Mind experience a significant level of violence, which surprises the Seeker.

What ideas have you developed regarding cultures of Ancient Mind (i.e., indigenous peoples)? Have you tended to view such cultures as essentially violent? Essentially peaceful? Both? What contributed to your development of these ideas?

THE BOOK OF MODERN MIND

10. The overarching Story of Modern Mind has at its core a sense of separation from the Web of Life, from the time humans began to settle in farming communities to the present day. Do you see any examples of this sense of separation expressing itself in your life? In the larger world? Throughout Modern Mind's history? If so, what are some of these examples?

11. The Sage asserts that the crises perpetrated by Modern Mind are *not* because humanity is fallen or flawed. Do you agree with the Sage's assertion? Why or why not?

THE GOSPEL OF PLANETARY MIND

12. The Sage asserts that Planetary Mind can emerge out of Ancient Mind and Modern Mind. How possible do you think this is? What do you think needs to happen to allow Planetary Mind to emerge?

13. In the overarching Story of Planetary Mind, the text refers to Planetary Mind awakening within the Holy Universe. What do you think the author means by the word "holy" in this context? Do you think this is an appropriate word to use in the context of the Sage's story? What does the word "holy" mean to you?

THE JOURNEY THROUGH
THE GREAT TRANSFORMATION

14. The Sage makes a distinction between the authority that is handed down from on high versus a shared collective authority and the internal authority within an individual. What are some examples that you've seen from each? What authority do you use, and how has that changed over your lifetime?

15. Of the three major crises driving the Great Transformation—ecological, social, and spiritual—which one has been paramount in your own awareness, and why? What parallels can you draw between this particular crisis and the others; i.e., how do you see them as related?

16. The Sage uses the story of how life responded to the oxygen catastrophe early in Earth's history as a metaphor for our time, that the catastrophes we face may drive an evolutionary change, or, as the Sage put it, the Great Transformation. How do you think the Great Transformation might play out as these catastrophes become larger?

STORIES OF MODERN MIND

17. The Sage asserts that examining issues of privilege is as important as any work to bring forth Planetary Mind. Have you had experience with doing this sort of work? How did it change your perceptions of privilege?
Note: before further exploring issues of privilege within a group setting, it is strongly recommended that a skilled and experienced facilitator be brought in to lead the discussion.

18. Do you agree with the Sage's assertion that the Universe is neither friendly nor unfriendly? If so, why? If not, which do you think it is, and why?

PERSONAL STORY

19. The Sage describes how our stories influence the "river of mind." What are some positive and negative stories from your life that influence your own river of mind? How have they helped or hindered you in your life? What stories do you want to change?

20. The Sage presents the Seeker with a model that outlines "Four Ways of Working in the World," which roughly coincide with (1) the philosopher, nun, or monk, (2) the helper or healer, (3) the activist, and (4) the storyteller. Which of these is most reflective of your own work in the world? What ways might you want to change in how you work in the world? What is your work in the world, and how has it changed over time? How have you discovered what your work in the world is?

CONNECTING WITH THE INFINITE

21. Ponder the fact that the atoms and molecules in your body hearken back to the Beginning of All Beginnings. What thoughts and feelings arise within you as you consider this? Do you think it is a useful practice to ponder such things? What other habits and practices do you think would be useful in connecting with the Infinite?

22. Have you ever tempered your expressions of awe toward creation? When and where did this happen? Similarly, when, if ever, have you been ridiculed for expressing awe at the beauty of the Web of Life or the Universe? How did you react?

EPILOGUE

23. In the Epilogue, the Seeker begins cleaning up the creek near his house in response to the serious disease with which he's been diagnosed. Has your life been affected by serious personal crises? If so, what have you learned from them, and what changes did they catalyze?

24. In the text of the Epilogue, the author refers to the Seeker "helping the creek heal itself," whereas in earlier drafts the text said, "restoring the creek as best he could." Why do you think this change was made? What difference in worldviews do you see between these two statements?

25. How might the story of the Seeker's illness be a metaphor for what humanity is going through now? What are some parallels between what the Seeker is going through versus what humanity is going through?

OTHER QUESTIONS

26. Of the four overarching Stories (Ancient Mind, the two chapters of Modern Mind, and Planetary Mind) presented in Part I, which one most closely reflects your own story and why?

27. There are only two places in the book where Earth is referred to as "*the* earth": in the two chapters of the overarching Story of Modern Mind. Why do you think the author might have chosen this construction?

28. Some readers of early manuscripts suggested that both the gender and the ethnicity of the Sage should be identified immediately rather than later in the text. Do you agree? What bias might these suggestions reveal? Why might the author have chosen not to do so?

29. What is the overarching Story that has guided your life? How has it changed over time? What stories do you feel have served you; what stories have hindered you? What stories do you want to change? Have you ever had difficulty in changing your stories? How hard do you think it is for people to change their stories?

 To learn about opportunities for discussing these and other questions with your fellow readers, visit www.theholyuniverse.com (scanning this code will also take you to the online discussion).

NOTES

In addition to giving credit for direct quotes I've used in this work, as well as sources for most of the numbers I present in the dialogues, I have also included selected annotations that are particularly compelling for me.

26 *we are the stories we tell ourselves:* Kapur, 2009.

55 *something other than chance . . . driving the Universe:* Wilber, 1996, p. 26. Wilber also relates a computation showing that the proverbial thousand monkeys randomly typing would create the works of Shakespeare once every 1,000,000,000 billion years or so—but the Universe created the galaxies, Earth, monkeys, Shakespeare, and typewriters in only 13.8 billion years. Using more manageable time scales, it is as if the monkeys would produce the works once every 730,000 centuries, and they did it in only one year . . . and only *after* everything else was created. He cites further calculations showing that "twelve billion years isn't even enough [time] to produce a *single enzyme* by chance [emphasis his]." See also Lovelock, 1979, p. 10, where he recalls his 1967 thoughts: "The climate and the chemical properties of the Earth now and throughout history seem always to have been optimal for life. For this to have happened by chance is as unlikely as to survive unscathed a drive blindfold through rush-hour traffic."

On the other hand, see also Christian, 2011, p. 100, where he relates Cesare Emiliani's assertion that if a "rule" (analogous to the laws of physics) locks in the correct letter (or a successful strategy for persisting) as it is typed, monkeys could randomly type the entire Bible within only a decade (after, of course, the creation of galaxies, stars, Earth, etc.). With such rules—and their mysterious origins—chance and randomness give way to a profound proclivity for powerful and rapid self-organization.

58 *The galaxies gave rise to the stars . . . gave rise to life:* Rogin, 2009.

86 *the Universe in the form of a human:* Swimme, 2009.

94 *Yet this particular society . . . lived with such levels of violence:* Many of us raised within the guise of Modern Mind hold romantic, utopian notions of indigenous peoples, but the story is more complicated. Some are indeed peaceful, others quite violent (see Plotkin, 1993, p. 245). During my own very brief trip to the rainforest, I visited one particular nation of indigenous people that had left behind its violent ways after being guided (and then choosing) to make contact with the modern world. I met the leader of this nation

who was indeed warm and welcoming—though it was quite sobering to learn later on that, given his age, this man was old enough to have likely participated in killing many other men in the time before his people made contact with the modern world.

95 *founding documents . . . created by the Haudenosaunee:* See Nelson, 2008, p. 61. The contribution of the Haudenosaunee (or Iroquois) was so fundamental to the forming of the U.S. Constitution that in 1987, on the 200th anniversary of the adoption of the Constitution, Congress formally acknowledged the contribution of and debt owed to the Haudenosaunee. See Congressional Record, 1987.

118 *All of these characteristics . . . describe Modern Mind to an astonishing degree:* The idea of humanity being in its adolescence has been put forth by a number of thinkers (e.g., see Elgin, 2009, pp. 161–162, and Renesch, 2012)—and refuted by others (see Plotkin, 2008, p. 7). Still, I find it a useful metaphor, especially when it is applied not to humanity as a species but instead toward its civilizations and societies.

119 *researcher . . . of a certain species of ants:* See Gordon, 1999, p. 21.

125 *history followed different courses . . . not because of biological differences among peoples:* Diamond, 1999, p. 25.

128 *There was once a village deep in a forest:* This story is based on a talk by Mark Plotkin for the Institute of Noetic Sciences. See Plotkin, 2006.

134 *the liberation of Modern Mind is bound up with its own liberation:* This text is adapted from the words of the artist, activist, and academic Lilla Watson, who reportedly prefers that the original quote, "If you have come here to help me, you are wasting your time. But if you have come because your liberation is bound up with mine, then let us work together," be attributed to "Aboriginal activists group, Queensland, 1970s." See "Lilla Watson," n.d.

139 *coming together of the two Minds . . . ancient prophecies:* See Perkins, 2004, p. 209, where the author discusses "The Prophecy of the Eagle and the Condor" and similar prophecies he encountered across cultures. See also Rogin, 2001.

157 *suffer from mental distress:* According to a 2008 report disseminated by the National Institute of Mental Health, one in four adults in the United States suffers from a diagnosable mental disorder in a given year. See Kessler, 2008.

159 *half of all species . . . extinct within a generation:* See "Current Mass Extinction," 2008.

159 *the sacred, talismanic . . . will not return:* Ulannsey, 2011.

164 *using one and a half times what Earth is capable of providing:* See Ewing, 2010, p. 9.

178 *millions of your as-yet-unknown fellow seekers:* see Ray, 2000, p. 39, where he and his co-author describe how most of the people they refer to as the "Cultural Creatives" think that "their worldview, values, lifestyle, and goals for the future are shared by only a few of their friends. They have little notion that there are 50 million of them. They do not know that they have the potential to shape the life of twenty-first century America." As of 2011, Ray has since increased his estimate of the population of Cultural Creatives in the U.S. from 50 million to 70 million.

180 *millions of organizations:* See Hawken, 2008, p. 2.

180 *two-fifths of the white population:* Robert M. Calhoun in Greene, 1991, p. 247.

182 *persuade people . . . horrendous acts:* The war criminal Hermann Goering once said, "Of course, the people don't want war . . . [but] it's a simple matter to drag the people along, whether it's a democracy or a fascist dictatorship or a Parliament or a communist dictatorship . . . All you have to do is tell them they are being attacked and denounce the pacifists for lack of patriotism and exposing the country to danger. It works the same way in any country." Gilbert, 1947, pp. 278-279.

193 *more like a picture of your own soul:* Alexander, 2004, p. 30.

200 *many are called, but few are chosen: King James Bible,* Matthew 22.14.

223 *we must daily break the body and shed the blood:* Berry, 1981, p. 281.

230 *whatever humanity does to the Web of Life:* This text is adapted from a piece attributed to Chief Seattle, but there is no actual record of Seattle's speeches. Most versions of this particular line can be traced back to an article written in the 19th century reporting on one of Seattle's speeches. See Smith, 1887, p. 3.

243 *every technology . . . social and political consequences:* See Mander, 1991, pp. 35-36. Mander offers an excellent critique of our unexamined acceptance of technological advances.

290 *throw off addictions . . . impacts that last for millennia:* Meadows, 2008, p. 165.

298 *upper two boxes . . . represent the group:* This model is based loosely (*very* loosely) on Ken Wilber's "Four Quadrant" model of what he defines as holarchies. See Wilber, 1996, pp. 73-82.

310 *All this is suffering:* "Sabbam idam dukkham" (from the early discourses attributed to the Buddha), which is often mistranslated as "Life is suffering." See Kalupahana, 1992, p. 86, and Harvey, 1990, pp. 47-53.

310 *Each second . . . four million tons:* See Swimme, 1996, p. 39.

322 *Turning Toward Evening:* This section is inspired by exercises outlined by Brian Swimme. See Swimme, 1996, pp. 26-28 and 47-53.

342 *Epilogue:* The epilogue is inspired by the story of a hero of mine whom I never got to meet: John Beal, a Vietnam veteran, who was "told that he had less than four months to live and advised by his doctor to find a hobby to take his mind off his pain and suffering." The hobby that came to him was the cleanup of a creek near his home. This work expanded to the restoration of the lower Duwamish River watershed in industrial Seattle, which he continued to steward until he did indeed die—twenty-seven years later. See Lyman, 2003, p. 36, and Langston, 2006.

SOURCES

While this isn't a scholarly work, it seemed fit to list some of the works—beyond those I've referenced in the Notes—that have influenced me in the creation of this story.

Abdullah, Sharif. *Creating a World That Works for All*. San Francisco: Berrett-Koehler Publishers, Inc., 1999. Print.

Abrams, Nancy Ellen, and Joel R. Primack. *The New Universe and the Human Future: How a Shared Cosmology Could Transform the World*. New Haven: Yale University Press, 2011. Print.

Alexander, Christopher. "Sustainability and Morphogenesis: The Birth of a Living World." *2004 Bristol Schumacher Lectures on Spirit, Nature, Matter*. Bristol, UK, 30 Oct. 2004. Berkeley: Center for Environmental Structure, 2004. PDF.

Alexie, Sherman. *Face*. Brooklyn: Hanging Loose Press, 2009. Print.

———. *The Toughest Indian in the World*. New York: Atlantic Monthly Press, 2000. Print.

Atlee, Tom. *Reflections on Evolutionary Activism: Essays, Poems, and Prayers from an Emerging Field of Sacred Social Change*. Eugene, OR: Evolutionary Action Press, 2009. Print.

———. Telephone interviews. 12 and 13 Aug., 2011.

Awakening the Dreamer, Changing the Dream Symposium. San Francisco: The Pachamama Alliance and Boulder: Gaiam Americas, Inc., 2011. DVD.

Barlow, Connie. *The Great Story: The 14 Billion Year Epic of Cosmos, Earth, Life, and Humanity Told in Meaningful, Inspiring Ways*. N.p., n.d. Web. 18 Sep. 2012. http://www.thegreatstory.org.

Benyus, Janine M. *Biomimicry: Innovation Inspired by Nature*. New York: William Morrow, 1997. Print.

Berry, Thomas. *The Dream of the Earth*. San Francisco: Sierra Club Books, 1990. Print.

Berry, Wendell. *The Gift of Good Land*. San Francisco: North Point Press, 1981. Print.

Brown, Cynthia Stokes. *Big History: From the Big Bang to the Present*. New York: The New Press, 2007. Print.

Bryson, Bill. *A Short History of Nearly Everything*. New York: Broadway Books, 2003. Print.

Cajete, Gregory. *Native Science: Natural Laws of Interdependence*. Santa Fe: Clear Light Publishers, 2000. Print.

Capra, Fritjof. *The Web of Life: A New Scientific Understanding of Living Systems*. New York: Anchor Books, 1996. Print.

Chaisson, Eric. *The Epic of Evolution*. New York: Columbia University Press, 2006. Print.

Christian, David. *Big History: The Big Bang, Life on Earth, and the Rise of Humanity*. Chantilly, VA: The Teaching Company, 2008. DVD.

———. *The Maps of Time: An Introduction to Big History*. Berkeley and Los Angeles: University of California Press, 2011. Print.

Churchill, Ward. *Struggle for the Land: Native North American Resistance to Genocide, Ecocide, and Colonization*. San Francisco: City Lights Books, 2002. Print.

Congressional Record. 16 Sep. 1987: Vol. 133, no. 140. Web. 28 May 2012.

"Current Mass Extinction Spurs Major Study of Which Plants to Save." *Office of Public Affairs, UC Santa Barbara*. University of California, Santa Barbara, 20 Oct. 2008. Web. 16 Sep. 2012.

Deming, Alison H., and Lauret E. Savoy, eds. *Colors of Nature: Culture, Identity, and the Natural World*. Minneapolis: Milkweed Editions, 2011. Print.

DeWolf, Thomas Norman. *Inheriting the Trade: A Northern Family Confronts Its Legacy as the Largest Slave-Trading Dynasty in U.S. History*. Boston: Beacon Press, 2008. Print.

Diamond, Jared. *Collapse: How Societies Choose to Fail or Succeed*. New York: Viking Penguin, 2005. Print.

———. *Guns, Germs, and Steel: the Fates of Human Societies*. New York: W. W. Norton, 1999. Print.

Dowd, Michael. *Thank God for Evolution: How the Marriage of Science and Religion Will Transform Your Life and Our World*. Tulsa, OK: Council Oak Books, 2007. Print.

Elgin, Duane. *The Living Universe: Where Are We? Who Are We? Where Are We Going?* San Francisco: Berrett-Koehler, 2009. Print.

———. *Voluntary Simplicity: Toward a Way of Life That Is Outwardly Simple, Inwardly Rich*. 2nd ed. New York: HarperCollins, 2010. Print.

Ewing, B., D. Moore, S. Goldfinger, A. Oursler, A. Reed, and M. Wackernagel. *The Ecological Footprint Atlas 2010*. Oakland, CA: Global Footprint Network, 2010. Print.

Garrow, David J. *Bearing the Cross: Martin Luther King, Jr., and the Southern Christian Leadership Conference*. New York: William Morrow and Company, Inc., 1986. Print.

Gilbert, G. M., *Nuremberg Diary*. New York: Farrar, Straus and Company, 1947. Print.

Gilman, Daniel. *Stumbling on Happiness*. New York: Vintage Books, 2007. Print.

Gordon, Deborah M. *Ants at Work: How an Insect Society Is Organized*. New York: The Free Press, 1999. Print.

Greene, Jack P., and J.R. Pole. *The Blackwell Encyclopedia of the American Revolution*. Cambridge, MA: Basil Blackwell, Inc., 1991. Print.

Harner, Michael J. *The Jivaro*. Berkeley and Los Angeles: University of California Press, 1972. Print.

Harvey, Peter. *An Introduction to Buddhism: Teachings, History and Practices*. New York: Cambridge University Press, 1990. Print.

Hawken, Paul. *Blessed Unrest: How the Largest Social Movement in History Is Restoring Grace, Justice, and Beauty to the World*. New York: Penguin Books, 2008. Print.

Heinberg, Richard. *The End of Growth: Adapting to Our New Economic Reality*. 2nd ed. Gabriola Island, B.C.: New Society Publishers, 2011. Print.

————. *The Party's Over: Oil, War, and the Fate of Industrial Societies*. 2nd ed. Gabriola Island, B.C.: New Society Publishers, 2005. Print.

Helfand, Judy, and Laurie Lippin. *Unraveling Whiteness: Tools for the Journey*. 2nd ed. Dubuque, IA: Kendall/Hunt Publishing Co., 2009. Print.

Holland, John Henry. *Emergence: From Chaos to Order*. Redding, MA: Addison-Wesley, 1998. Print.

Hopkins, Rob. *The Transition Companion: Making Your Community More Resilient in Uncertain Times*. White River Junction, VT: Chelsea Green Pub., 2011. Print.

Johnson, Steven. *Emergence: The Connected Lives of Ants, Brains, Cities, and Software*. New York: Scribner, 2001. Print.

Kalupahana, David J. *A History of Buddhist Philosophy: Continuities and Discontinuities*. Honolulu: University of Hawaii Press, 1992. Print.

Kane, Joe. *Savages*. New York: Vintage Books, 1996. Print.

Kapur, Shekhar. *We Are the Stories We Tell Ourselves*. TEDIndia Conference, November 2009. Web. 5 Apr. 2010.

Kauffman, Stuart. *Reinventing the Sacred: A New View of Science, Reason, and Religion*. New York: Basic Books, 2008. Print.

Kessler, R. C., W. T. Chiu, O. Demler, and E. E. Walters. "Prevalence, Severity, and Comorbidity of Twelve-Month DSM-IV Disorders in the National Comorbidity Survey Replication (NCS-R)." *Archives of General Psychiatry*, 2005 Jun.; 62 (6): pp. 617-27, as quoted in "The Numbers Count: Mental Disorders in America." *National Institute of Mental Health*, n.d. Web. 28 May 2012.

Kirsch, Jonathan. *The Harlot by the Side of the Road: Forbidden Tales of the Bible*. New York: Ballantine Books, 1997. Print.

Langston, Jennifer. "John Beal: 1950-2006: River Steward Never Backed Down." *Seattle PI*, 25 Jun. 2006. Web. 15 Jul. 2012.

Leakey, Richard E., and Roger Lewin. *The Sixth Extinction: Patterns of Life and the Future of Humankind*. New York: Doubleday, 1995. Print.

Liebes, Sidney, Elisabet Sahtouris, and Brian Swimme. *A Walk Through Time: From Stardust to Us: The Evolution of Life on Earth*. New York: Wiley, 1998. Print.

"Lilla Watson." Wikipedia, n.d. Web. 28 May 2012.

Lovelock, James. *The Ages of Gaia*. New York: Norton, 1988. Print.

———. *Gaia: A New Look at Life on Earth*. Oxford, New York: Oxford University Press, 1979. Print.

———. *The Vanishing Face of Gaia*. New York: Basic Books, 2009. Print.

Lyman, Francesca. "Restoring Nature, Restoring Yourself." *YES! Magazine*. Spr. 2003. Print.

Mander, Jerry. *The Capitalism Papers*. Berkeley: Counterpoint, 2012. Print.

———. *In the Absence of the Sacred: The Failure of Technology and the Survival of the Indian Nations*. San Francisco: Sierra Club Books, 1991. Print.

McKibben, Bill. *Deep Economy: The Wealth of Communities and the Durable Future*. New York: Times Books, 2007. Print.

Meadows, Donella. *Thinking in Systems: A Primer*. Diana Wright, ed. White River Junction, VT: Chelsea Green Publishing, 2008. Print.

Nelson, Melissa K., ed. *Operating Instructions: Indigenous Teachings for a Sustainable Future*. Rochester, VT: Bear & Company, 2008. Print.

The New Layman's Parallel Bible. Grand Rapids, MI: Zondervan Bible Publishers, 1981. Print.

Novak, Philip. *The World's Wisdom: Sacred Texts of the World's Religions*. San Francisco: HarperSanFrancisco, 1994. Print.

Perkins, John. *Confessions of an Economic Hit Man*. San Francisco: Berrett-Koehler Publishers, Inc., 2004. Print.

Plotkin, Bill. *Nature and the Human Soul: Cultivating Wholeness and Community in a Fragmented World*. Novato, CA: New World Library, 2008. Print.

Plotkin, Mark. "The Marriage of Indigenous Wisdom and Modern Technology to Preserve the Amazon." *Global Consciousness: The Missing Piece of the Sustainability Puzzle.* Institute of Noetic Sciences. Scottish Rites Temple, Oakland, 22 Apr. 2006. Presentation.

————. *Tales of a Shaman's Apprentice: An Ethnobotanist Searches for New Medicines in the Amazon Rain Forest.* New York: Viking Penguin, 1993. Print.

Primack, Joel R., and Nancy Ellen Abrams. *The View from the Center of the Universe: Discovering Our Extraordinary Place in the Cosmos.* New York: Riverhead Books, 2006. Print.

Quinn, Daniel. *Ishmael.* New York: Bantam/Turner Books, 1992. Print.

————. *The Story of B.* New York: Bantam Books, 1996. Print.

Radin, Dean. *Entangled Minds: Extrasensory Experiences in a Quantum Reality.* New York: Paraview Pocket Books, 2006. Print.

Ray, Paul H., and Sherry Ruth Anderson. *The Cultural Creatives: How 50 Million People Are Changing the World.* New York: Harmony Books, 2000. Print.

Renesch, John. *The Great Growing Up: Being Responsible for Humanity's Future.* Prescott, AZ: Hohm Press, 2012. Print.

Robin, Vicki, and Joe Dominquez, with Monique Tilford. *Your Money or Your Life: Nine Steps to Transforming Your Relationship with Money and Achieving Financial Independence.* Revised ed. New York: Penguin Books, 2008. Print.

Rogin, Neal, and Drew Dellinger. *The Awakening Universe.* San Francisco: The Pachamama Alliance, 2009. DVD.

————. *The Eagle and The Condor.* San Francisco: The Pachamama Alliance, 2001. DVD.

Seligman, Martin E. P. *Authentic Happiness: Using the New Positive Psychology to Realize Your Potential for Lasting Fulfillment.* New York: Free Press, 2002. Print.

Simons, Nina. *Moonrise: The Power of Women Leading from the Heart.* Rochester, VT: Park Street Press, 2010. Print.

Smith, H. A. "Early Reminiscences. Number Ten. Scraps From a Diary. Chief Seattle—A Gentleman by Instinct—His Native Eloquence. Etc., Etc." *Seattle Sunday Star*, October 29, 1887: 3. Print.

Spier, Fred. *Big History and the Future of Humanity.* West Sussex, UK: Blackwell Publishing Ltd., 2011. Print.

Stewart, John. *Evolution's Arrow: The Direction of Evolution and the Future of Humanity.* Canberra, ACT, Australia: The Chapman Press, 2000. Print.

Surowiecki, James. *The Wisdom of Crowds: Why the Many Are Smarter Than the Few and How Collective Wisdom Shapes Business, Economies, Societies, and Nations.* New York:

Doubleday, 2004. Print.

Swimme, Brian, interview in *The Awakening Universe*. San Francisco: The Pachamama Alliance, 2009. DVD.

———. *The Hidden Heart of the Cosmos: Humanity and the New Story*. Maryknoll, NY: Orbis Books, 1996. Print.

———. *The Universe Is a Green Dragon: A Cosmic Creation Story*. Santa Fe: Bear, 1985. Print.

———, and Thomas Berry. *The Universe Story: From the Primordial Flaring Forth in the Ecozoic Era—A Celebration of the Unfolding of the Cosmos*. San Francisco: Harper San Francisco, 1992. Print.

Ulansey, David, interview in "Where Are We?" *Awakening the Dreamer, Changing the Dream* Symposium. San Francisco: The Pachamama Alliance, and Boulder: Gaiam Americas, Inc. 2011. DVD.

Watts, Alan. *Buddhism, The Religion of No-Religion: The Edited Transcripts*. Rutland, VT: Charles E. Tuttle Co., Inc., 1996. Print.

———. *Eastern Wisdom, Modern Life: Collected Talks, 1960-1969*. Novato, CA: New World Library, 2006. Print.

———. *The Philosophies of Asia: The Edited Transcripts*. Rutland, VT: Charles E. Tuttle Co., Inc., 1995. Print.

Wilber, Ken. *A Brief History of Everything*. Boston: Shambhala Publications, Inc., 1996. Print.

Wolff, Robert. *Original Wisdom: Stories of an Ancient Way of Knowing*. Rochester, VT: Inner Traditions International, 2001. Print.